THE UPPER ATMOSPHERE

THE UPPER ATMOSPHERE

PART IV
OF
SOLAR-TERRESTRIAL PHYSICS/1970

COMPRISING THE PROCEEDINGS OF THE
INTERNATIONAL SYMPOSIUM ON SOLAR-TERRESTRIAL PHYSICS
HELD IN LENINGRAD, U.S.S.R.
12–19 MAY 1970

Sponsored by COSPAR, IAU, IUGG-IAGA, and URSI

S. A. BOWHILL
Editor

E. R. DYER
General Editor of the Proceedings

SPRINGER-SCIENCE+BUSINESS MEDIA, B.V.

ISBN 978-90-277-0213-5   ISBN 978-94-010-3132-5 (eBook)
DOI 10.1007/978-94-010-3132-5

TABLE OF CONTENTS

INTERPRETATION OF IONOSPHERIC EFFECTS OF
SOLAR FLARES

A. P. MITRA

National Physical Laboratory, New Delhi-12, India

1. Introduction

The classical methods of recording ionospheric effects due to solar flares include: sudden absorption increases observed with HF commercial CW transmissions (SWF) or with riometers (SCNAs), and with pulse absorption techniques; VLF/LF phase and amplitude changes (SPA, SEA, SES); and sudden frequency deviations observed with HF standard frequency transmissions (SFD). Figure 1 gives an example of some of these effects recorded during the flare of January 30, 1968.

These effects are, however, indirect ones; these arise because of an enhancement in ionization which progressively decreases with height from a factor of 5–10 around 70–80 km to about 50–100% in the E region and about 1–20% in the F region. Figure 2 gives a rough picture of the different degrees of ionization enhancement at different levels. Attempts to determine these ionization changes quantitatively are relatively more limited; amongst the more important efforts are those by the Pennsylvania State University which uses a high-power wave-interaction technique and obtains quick-run profiles of the D-region ionization during the entire course of the flare (Rowe *et al.*, 1970); those by Belrose and his colleagues who use a partial-reflection technique (Belrose, 1969); a remarkable series of profile determinations from 100 to about 300 km by the incoherent-scatter equipment at Arecibo for the two large flares occurring on May 21 and 23, 1967 (Thome and Wagner, 1967); and two series of rocket flights by Somayajulu and Aikin (1969, 1970) into the flares occurring on January 15, 1968 and August 21, 1968. Since such works are necessarily limited, there have been several attempts to use the more conventional SID technique for profile studies. This has been done by May (1966) for the flare of October 7, 1948 with VLF observations and by Deshpande and Mitra (1970) with multifrequency SCNA observations.

In any physical study of flare-associated ionospheric effects, it is desirable to have additionally the following information:

(1) The complete time history and the spectral distribution of the ionizing flux, along with *changes* in the spectral distribution.

(2) The nature and concentrations of the atmospheric constituents ionized, and

(3) The nature of the effective loss rate.

It is, however, very rare to have all this information for any one event. Firstly, while entire time histories are now recorded for one or more bands (e.g. 0–3 Å, 0–8 Å, 8–20 Å, 44–60 Å, 2–12 Å), it is not easy to build a reliable spectral distribution out of these measurements, partly because the detector response itself is a function of

Dyer (ed.), Solar-Terrestrial Physics/1970: Part IV, 1–26. All Rights Reserved.
Copyright © 1972 by D. Reidel Publishing Company.

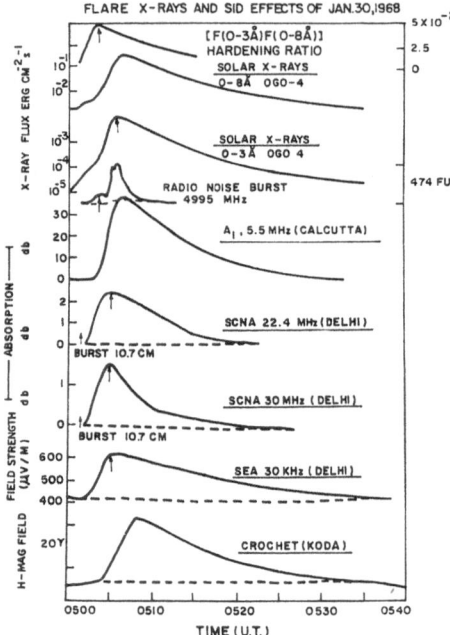

Fig. 1.   Examples of ground-based conventional SIDs recorded during the solar flare of January 30, 1968, shown along with solar X-rays in the bands 0–3 Å and 0–8 Å monitored by OGO-4 satellite. The hardening ratio $F(0\text{–}3 \text{ Å})/F(0\text{–}8 \text{ Å})$ is also plotted. Note that the maximum of the SIDs occur during this event at the time of maximum hardening of X-rays and not at the time of maximum X-ray flux either at 0–3 Å or 0–8 Å (after Deshpande *et al.*, 1970).

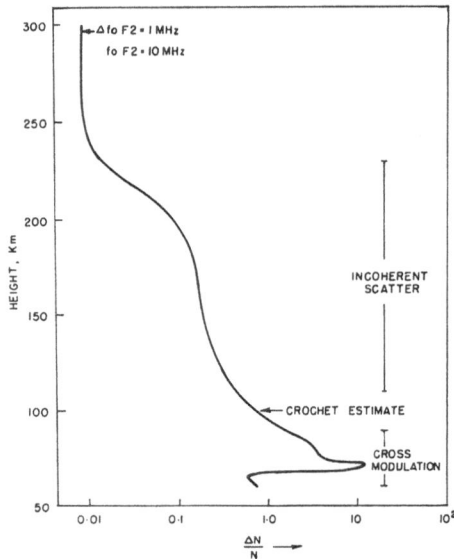

Fig. 2.   Representative diagram showing ionization enhancement at various levels in the ionosphere during a moderate solar flare. Sources of information for different heights are indicated.

the spectral temperature and partly because this temperature is not constant over the band. Secondly, difficulties arise specially in the D region where much of the pre-flare ionization is controlled by minor constituents (such as NO) and the relative contributions of the different ionization sources are uncertain. Thirdly, evidence exists to show that there are changes in the D region loss rates during a flare and that these changes may well vary from flare to flare; however, at heights above 100 km there is no evidence for any change in the loss rate.

2. SID Phenomenology

Much work has been done relating SIDs with Hα flares. In contrast, comparisons with X-rays or with solar radio bursts (which are more relevant) have only been made relatively recently with increasing availability of satellite measurements of solar X-rays and increased monitoring of solar radio bursts over a wide range of frequencies. X-rays below 20 Å are now monitored extensively and continuously in a number of bands. In addition, several EUV lines (e.g. He$_{II}$ 303.8 Å, Fe$_{XV}$ 284.1 Å, Fe$_{XVI}$ 331.3 Å) have shown enhancements during flares.

Many relationships between SIDs, solar X rays, radio bursts and Hα emissions have been identified. Many of these are merely statistical in nature. We point out in this section only some of the more important features:

(1) A curious feature on the appearance of SIDs is that if two consecutive periods were chosen so that the 10 cm flux level was lower than average in one and greater than average in the other, but there was an abundance of Hα flares in both, then the SIDs are abundant in the latter, but nearly or entirely absent in the former. An example is given in Figure 3. While this example is for the period of the IGY, the same situation was found to exist during the IQSY.

(2) A question that one nearly always asks is to what extent the SIDs can be used to indicate the characteristics of other flare-associated events. The following are some of the more important results:

(a) Percentage association of SIDs with the X-rays increases with the energy of the X-ray band. It is 60–70% for soft X-rays, but 90% for 10–50 keV X-rays.

(b) The highest percentage of occurrence occurs apparently with SPA and SWF; these are then the most sensitive indicators of soft X-rays. When X-ray flares concurrent with Hα and radio noise bursts are considered, *all types of SIDs* including SCNA, SFD and crochet occur in about 80% of the cases.

(c) The capability of an X-ray flare to induce an SID effect depends on the flux level as well as on the spectral composition. The threshold flux for 0–8 Å band is 1–2×10^{-3} erg cm^{-2} sec^{-1} ($T = 2 \times 10^{6}$ K), provided the hardening ratio is at least 1.5×10^{-2}. Smaller fluxes can produce SIDs if the spectral hardening increases the proportion of the X-ray energy below 3 Å to about 13% of total enhancements in 0–8 Å flux. Conversely, large enhancements in 0–8 Å flux are capable of producing SIDs without much hardening. (Figure 4).

(d) X-ray flares associated with most of the complex cm radio bursts (80%) and

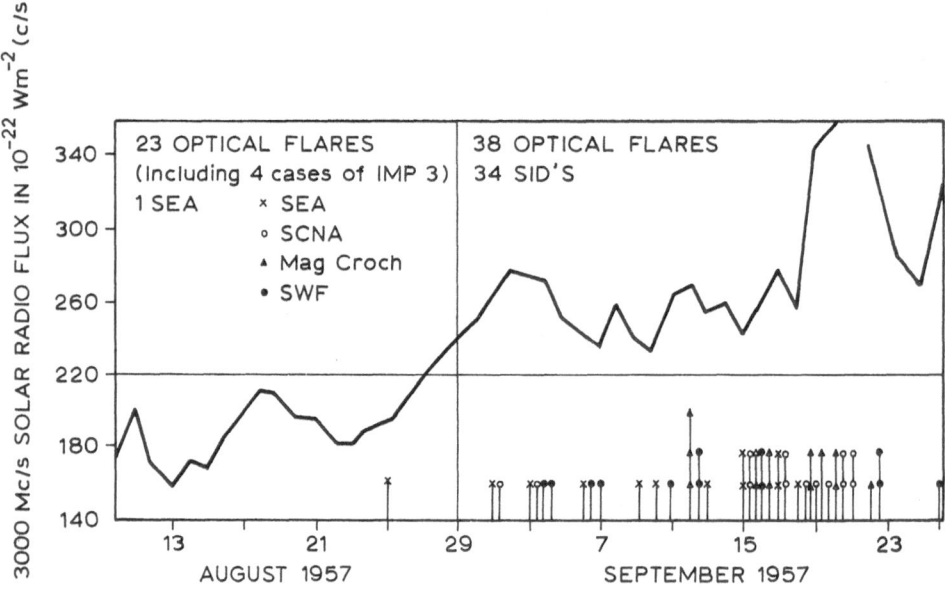

Fig. 3.   Occurrence of SIDs during two successive intervals in which 10.8 cm solar radio flux was
widely different, but optical flares were unusually abundant (after Mitra *et al.*, 1964).

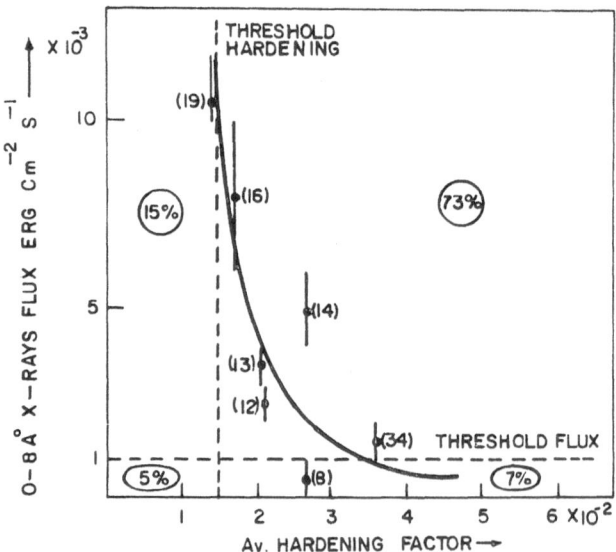

Fig. 4.   Average hardening factor and the threshold X-ray flux in 0–8 Å band (after
Deshpande *et al.*, 1970).

with impulsive and GRF bursts of size above 60 and 20 flux units ($\times 10^{-22}$ Wm^{-2} Hz^{-1}) respectively, invariably produce an SID effect.

The X-ray flux in 0–3 Å band peaks earlier than in 0–8 Å (Kreplin *et al.*, 1969). However, the VLF-LF effects believed to be mainly controlled by 0–3 Å band reach their maximum later than HF SID effects like SCNA and SWF.

The relaxation time of SID effects with reference to soft X-ray flares is about 2–3 min (Figure 5). SCNA and to some extent SWF tend to show smaller relaxation time. The relaxation time indicates the values of ($\alpha_{eff}N$) as $(25–40) \times 10^{-4}$ sec^{-1}.

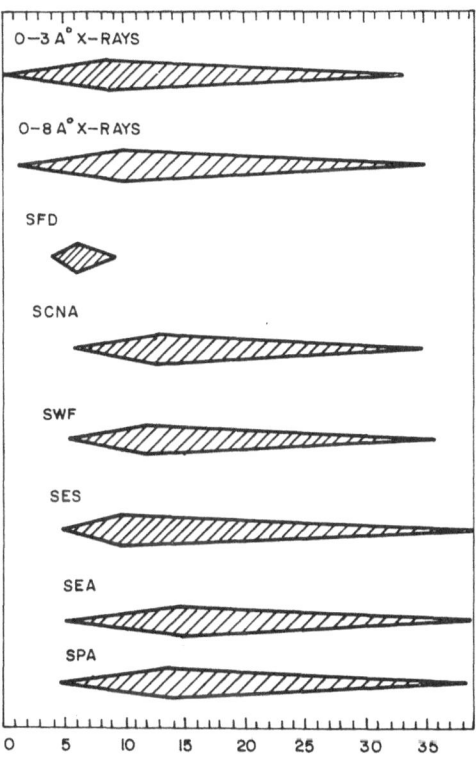

Fig. 5.   Average times of growth and decay for different SIDs compared with those of X-rays in 0–3 Å and 0–8 Å bands.

(3) Sakurai (1968) has discovered that when one observes SEAs at 10, 21 and 27 kHz, one can identify three separate types of events, each associated with specific types of cosmic ray events. This is shown in Figure 6. Type A is one in which there is a sudden enhancement in all the three frequencies; type B is one in which 10 kHz is unaffected, while there is enhancement at 27 and 30 kHz, and type C is one in which there is a decrease in intensity at 10 kHz. Sakurai finds that SEAs associated with cosmic rays belong mainly to Type C. Those of type sudden-C are mainly associated with BeV cosmic ray particles, while the SEAs of type slow-C are accompanied by MeV cosmic ray flares of F and F* types (Table I).

TABLE I

Solar cosmic-ray flares and associated SIDs

Cosmic-ray flare	I	II
Cosmic rays	GeV particles (UI or SI) and MeV particles of type F and F* (PCA)	MeV particles of type S (PCA)
Spectra of type IV burst	F↑ IVμ IVm → freq.	F↑ IVm IVμ → freq.
SID		
SWF (type IV spectra)	Sudden drop-out	Slow drop-out
SEA 10 kHz	Decrease	Increase or invariable
21 kHz	Increase	Increase
27 kHz	Type C (sudden or slow C)	Types A and B
SFD	Yes	No
f_0F_2	Increase	No

(4) When riometers are operated at more than one frequency, one can identify from a change in the frequency law as the flare progresses, events in which the X-ray spectra have been usually hard. Examples are given in Figure 7. Under normal circumstances, absorption at two frequencies is related by the equation

$$\frac{A(f_1)}{A(f_2)} = \frac{(\omega_2 \pm \omega_L)^2 + v^2}{(\omega_1 \pm \omega_L)^2 + v^2}. \tag{1}$$

As the X-ray spectrum hardens, ionization is produced at increasingly lower levels where v^2 begins to be comparable to $(\omega \pm \omega_L)^2$, bringing the frequency exponent down from its pre-flare value of -2 to generally around -1.5, and in extremely rare cases to around -1.0. The information, however, is only qualitative; quantitative evaluation of the X-ray spectrum of the decreasing level of peak absorption involves many assumptions.

(5) The SID time curves are, in general, similar to those in X-rays or in the centimeter radio burst. Slow and impulsive radio or X-ray events produce correspondingly slow and impulsive SIDs. There are, however, differences in detail. The soft X-ray (e.g. $\lambda\lambda$ 0–8 Å, 2–12 Å) enhancement begins before the SID and continues even after the end of the SID; 10–50 keV X-rays, however, correspond closely with the time of start and rate of growth of X-rays. There is some indication – this is clear in the examples of January 30, 1968 which we show in Figure 1 – that the SID follows the time change of the *spectral composition*, rather than the time development of X-ray flux (Deshpande *et al.*, 1970).

Comparison with radio noise bursts shows that when bursts occur with a clear single peak, not superimposed on GRF, the SID shows a sudden onset and a rapid rate of growth. When, however, the impulsive cm radio burst is superimposed on a gradual rise and fall, the SIDs correlate better with the GRF in start, in growth, as well as in later development. The impulsive burst has little connection with SWF.

(6) For the SFDs, the only F-region effect routinely recorded and, consequently,

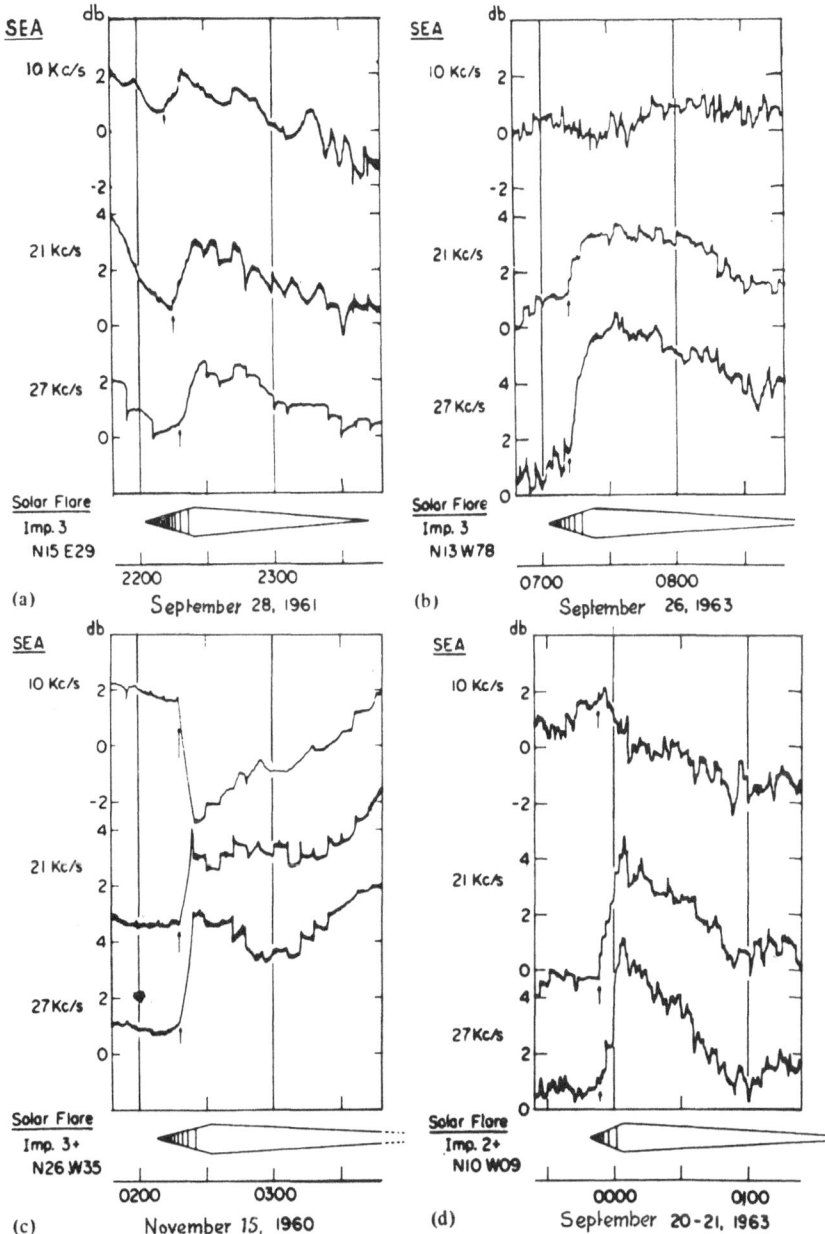

Fig. 6. Examples of different types of SEAs (see Table I) associated with different types of solar cosmic ray flares (after Sakurai, 1968).

the only EUV routine ionospheric monitoring, Donnelly (1969) has found that flare flashes of some EUV lines (e.g. He 303.4 Å, O v 629.7 Å, HLy 972.5 Å, C III 977 Å and H Iα 1215.7 Å) have time dependence in close agreement with the total radiation responsible for SFDs, except during the negative decay phase, but others (e.g. Fe xv 284.1 Å, Fe xvi 325.3 Å, Si xii 499.3 Å, Hg x 625.3 Å and Ni viii 770.4 Å, which are normally coronal lines) have a much shorter time dependence than the radiation mainly responsible for SFDs.

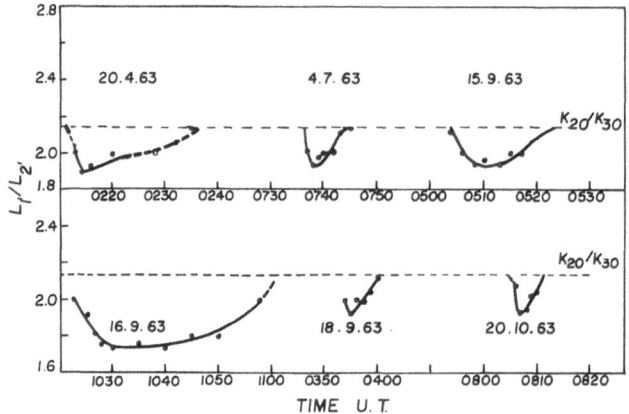

Fig. 7.   Examples of riometer flare events indicating unusual hardening in soft X-rays.

3. D Region During Flares

SIDs originating in the D region are abundant. These include: SWF and SCNA, SPA, SEA and SES. Most observations, however, provide only qualitative information on the D region. As mentioned before, quantitative information is beginning to become available from experiments with partial reflection, cross modulation, VLF amplitude and phase measurements, multifrequency absorption measurements and from direct rocket soundings.

3.1. IONIZATION CHANGES

Appropriate estimates of average D-region ionization changes have been obtained in the past from measurements of SCNAs or of A1 absorption and from the magnitudes of crochets. Table II gives a summary of these results after Mitra (1968). A more recent work by Deshpande *et al.* (1970), relating increases in absorption with relative enhancement P/Q in the 2–12 Å X-ray band shows that the rate of increase in absorption is very rapid for small values of P/Q, but is rather slow for large X-ray enhancements ($P/Q=20$).

The important thing, however, is to find a complete electron-density profile for the *same flare* and for the *same place*. The methods so far used and the flares examined are summarized in Table III.

The high-power wave interaction technique of Penn State appears to have been

TABLE II

Summary of SID observations

Phenomena	Method	Source	SID effects			Parameter	Relaxation time (min)		Remarks
			Class 1	Class 2	Class 3		τ_0	τ_f	
F.O.	4 Mc/sec	Appleton and Piggott (1954)		Between 5 and 10			—	—	
	SCA (18.3 Mc/sec)	Shain and Mitra (1954)	2.5	4.5	7.0	A_f/A_0	27	30	
	SCA (22.4 Mc/sec)	Mitra and Sarada (unpublished)		Between 3 and 35			35	—	Effect corresponding to flare causing absorption increase by a factor of 3
Low level reflection	2 Mc/sec	Gardner (1959)	Excess ionization of about 1000/cm³ at 60–76 km; $N_f/N_0 \approx 20$			—	—	—	
SPA	16 kc/sec	Bracewell and Straker (1949)	2	5.5	9.0		23	8 min (class 2) 3 min (class 3)	
	40–113 kc/sec	Weeks and Stuart (1951)		Same		Δh (km)	—	—	
	75 kc/sec	Houston et al. (1957)		2 to 4			16	3	$\Delta h = 5$–8 km
SEA	2385 kc/sec	Findlay (1951) Ellison (1950)		1.26			—	—	$\Delta h' \doteqdot \Delta h$
	27 kc/sec	Sachdev (1958)					21	7	
Crochet		From McIntosh (1951) data	0.50	0.70	1.30	$\Delta H/H$			

TABLE III

Flare time D-region electron density profiles

Method	Authors	Height range (km)	Flare events recorded	Remarks
Cross-modulation	1. Rowe et al. (1970)	55–90	1. Oct. 21, 1968 2. Oct. 29, 1968 3. Jan. 17, 1969 4. Feb. 25, 1969 5. Feb. 27, 1969 6. Mar. 12, 1969 7. Mar. 17, 1969 8. Mar. 18, 1969 9. Mar. 19, 1969 10. Mar. 20, 1969 11. Mar. 21, 1969 12. Mar. 25, 1969 13. Mar. 26, 1969 16. Apr. 21, 1969 17. Apr. 30, 1969 18. May 29, 1969 19. Nov. 28, 1969	Complete profiles every 3–4 min observations. Coordinated with NRL satellite X-ray monitoring
	2. Penn. State (private communication)			
Partial reflection	Belrose and Cetiner (1962) Belrose (1969)	60–85	Mar. 1, 1962 Mar. 25, 1968	
LF/VLF phase and amplitude	May (1966) Burgess and Jones (1967)	60–85	Oct. 7, 1948	May's calculations based on Deeks profile, later modified by Bain and May
Multifrequency absorption (combined A1 and A2)	Deshpande et al. (1970)	60–85	Jan. 30, 1968	Used A2 observations on 20, 22.4 and 30 MHz at Delhi and 5.5 MHz A1 observations at Delhi
Rocket experiments	Somayajulu and Aikin (1970)	65–100	1. Jan. 16, 1968 2. Aug. 21, 1968 (3 rockets)	Launched during decay phase of A class 1 flare class 1B flare Flight 3 min after X-ray peak. Second and third flights 12 and 24 min later

the most successfully used so far. Since the installation of the present high-power equipment in November, 1967, soundings have been made during more than a dozen flares. The experiment produces profiles of N_e, over the range of 55–90 km, at the rate of about one profile every 3 or 4 min. Figure 8(a) shows the profiles obtained during the flare of 21 October 1968, occurring at about 1730 UT. This was a class 2B flare with a strong X-ray enhancement (about a factor of 50 below 8 Å). The normal profile is one obtained at about 1710 UT, just before the flare began. As the flare began, N_e at heights above 70 km began to increase without any appreciable change at and below 70 km (1727 UT profile). The peak effect was observed at 1739 UT; the profile at this time was quite different in shape from the normal one, suggesting a different production source during normal and disturbed times. During the decay phase the profile at 1801 UT was identical with that of 1727 UT at levels above 70 km,

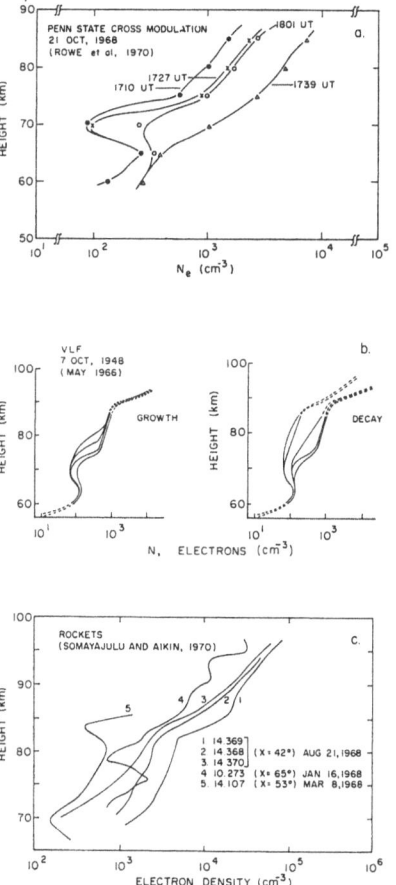

Fig. 8. Ionization profiles for the D region obtained during solar flares by different techniques: (a) Profiles obtained by Rowe et al. (1970) from Penn State cross modulation for the solar flare of October 21, 1968. (b) Those obtained by May (1966) from VLF measurements in Cambridge for the flare of October 7, 1948. (c) Those obtained by Somayajulu and Aikin (1970) with rocketborne instruments for the flares of January 16 and August 21, 1968.

but was considerably different below. Other flares examined behaved similarly; the
difference was in the degree of enhancement.

The second method in the table is that of partial reflection. Profile studies have
been published, however, only from the group of Belrose in Ottawa. While only two
profiles have been published so far, one for the flare of March 1, 1962 (Belrose and
Cetiner, 1962); and the other for the flare of March 25, 1968, many other events are
known to have been observed. These profiles are plotted along with the Penn State
profile of October 21, 1968 in Figure 9.

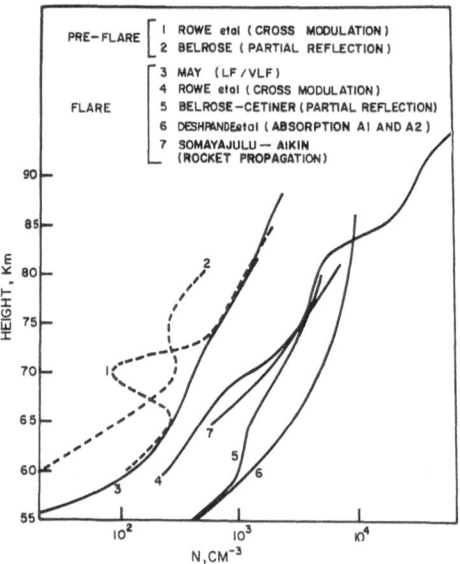

Fig. 9. Comparison of flaretime ionization profiles obtained by different authors using different
techniques. The profiles refer to different flare events, but all are in weak-to-moderate categories.
Pre-flare profiles for two events are given by the dotted curves.

The third method is the use of observations of the propagation of LF and VLF
radio waves. Using Cambridge observations of phase and amplitude variations in
the LF and VLF, May (1966) has examined two SIDs. The first is an 'average' SID
which gives rise to most of the observed characteristics of the propagation of radio
waves and the second is an SID which occurred on 7 October, 1948 during which
continuous measurements were made at Cambridge of the strength and phase of
downcoming radio waves at 16 and 70 kHz at steep incidence. The growth and decay
of ionization for the flare of 7 October 1948 is given in Figure 8c, and the profile
during the peak of the event is given in Figure 9. These profiles were, however, based
on Deek's normal N_e profile and not the revised profile later given by Bain and
May (1967) and are, to that extent, unreliable.

The next method shown is that of absorption. Deshpande and Mitra (1970) have
derived a number of profiles from: (i) SCNA measurements at a pair of frequencies,
(ii) simultaneous measurements of flare effects in A1 and A2 absorption, and

(iii) simultaneous use of SCNA at HF and SPA at VLF. Deshpande and Mitra have examined about a dozen flare events with this method, but the one that is of particular importance is the flare of January 30, 1968, for which, in addition to Dehli measurements of SCNA on 22.4 and 30 MHz, increase in absorption was also recorded at Calcutta at 5.5 MHz as a function of time and for the entire course of the flare event. The profile is shown along with other flare profiles in Figure 9.

The final method shown is that of direct rocket soundings. While its importance is obvious, the practical problems of proper timing of the launchings make this an exceedingly difficult experiment to perform. The only rocket firings made so far are by Somayajulu and Aikin (1969, 1970) who have successfully obtained electron-density profiles from two series of flare time launchings. The first series consisted of two rocket launchings during the decay phase of a class 1 flare that occurred on January 15, 1968. The second series were flown into a class 1B flare in August 1968, the first flight being near the peak of the flare, the second during the decay phase and the third at the end of the flare. The payloads were instrumented for the simultaneous determination of ionizing $L\alpha$ and X-rays along with the electron density. The development and decay of ionization are given in Figure 8a and the peak-flare profile obtained for the flare of August 21 is given in Figure 9.

It is interesting to compare the different profiles obtained by different techniques

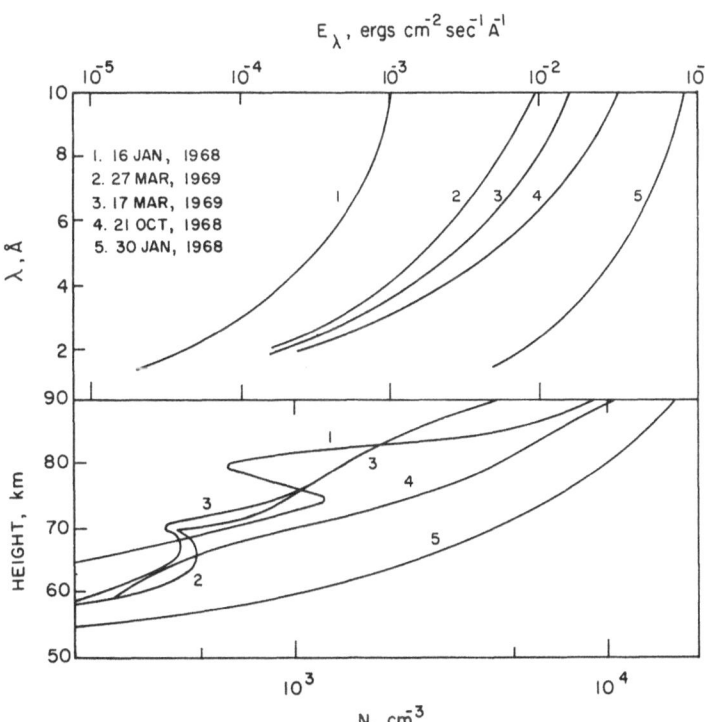

Fig. 10. X-ray spectral distributions in 2–8 Å for selected flare events along with the corresponding electron density profiles. The X-ray events are arranged in increasing order of flux values.

and for different flare events. We see that the flare profiles are generally similar in shape, excepting that some have larger enhancements than the others, and that in the case of Somayajulu-Aikin rocket profiles there is an unexpected trough at 80 km that is not seen in the other cases. When one arranges the flare events according to increasing flux values (Figure 10), we find that the electron-density profiles then line up in increasing order. These confirm the reliability and consistency of these measurements as well as the predominant role played by flare X-rays in the production of the additional ionization.

3.2. INFORMATION ON NITRIC OXIDE FROM SIMULTANEOUS MEASUREMENTS OF ELECTRON DENSITY AND SOLAR X-RAYS DURING FLARES

A source of much controversy is the magnitude of the non-X-ray production rate. The controversy rests principally on the concentration of nitric oxide for which Pearce's rocket results (1969) yield a disconcertingly large value of 10^9 cm^{-3} at 70 km (although more recent rocket measurements by L. G. Meira (private communication) yield values about an order of magnitude less) against the ionospheric estimates of 10^7 cm^{-3} obtained by Mitra (1969). The ease with which even a minor X-ray enhancement produces an SID is inconsistent with Pearce's NO values which would give, for $\chi = 50°$, q values around 50 cm^{-3} at both 70 and 80 km. These will be exceeded by solar X-rays only for large X-ray enhancements. It thus seems that an accurate measurement of the changes in N_e during a flare at any fixed level coupled with a knowledge of the X-ray production rate at that time can provide useful information on the upper limit of nitric oxide.

This has recently been attempted by Rowe et al. (1970) with the flare of October 21, 1968 shown in Figure 8a. Integrated fluxes I(0–3), I(0–8) and I(8–20 Å) were obtained by the NRL from the 0.5–3 Å, 1–8 Å and 8–12 Å photometers aboard the SOLRAD 9 satellite. Spectra were calculated for different times of the flare using a power law, and X-ray production rates were computed at these times. Figure 11 shows a time history of N_e and the X-ray production rate, q_x, at 80 km. If one interprets the results from the equation:

$$\frac{d\lambda}{dt} + \lambda^2 \left[\alpha_i N_e\right] + \lambda \left[(\alpha_i + \alpha_D) N_e + \frac{1}{N} \frac{dN_e}{dt}\right]$$
$$+ \left[\frac{1}{N_e}\left(\frac{dN_e}{dt} - q\right) + \alpha_D N_e\right] = 0 \qquad (2)$$

and retain the term $d\lambda/dt$ (since there is no compelling physical reason for taking $d\lambda/dt = 0$ during a flare), then one finds that q_1, the non-X-ray production term, depends on the pre-flare values of α_D, α_i and λ_0. The Penn State approach was to assume values for q_1, for the effective ion-ion recombination coefficient α_i, and for the pre-flare value of $\lambda(=\lambda_0)$. The equilibrium equation (with $d\lambda/dt = dN_e/dt = 0$) was then solved for α_D. The differential equation was then integrated numerically using the above values, and using time histories of N_e and q_x as shown in Figure 11.

$q_1 = 8$ cm^{-3} sec^{-1} is the largest q_1 which gives positive λ for all times. If q_1 is reduced below 3 cm^{-3} sec^{-1}, λ does not return to λ_0 in the same time interval that N_e returns to normal. The conclusion is that $q_1 < 10$ cm^{-3} sec^{-1} at 80 km if $\alpha_i \simeq 1 \times 10^{-7}$ cm^3/sec and $\lambda_0 < 8$; q_1 is probably about 3 or 4 cm^{-3} sec^{-1} if λ_0 is about 5 (Baker, 1969). If all of the q_1 at 80 km is due to Lα ionizing NO then $n(NO) \simeq 3 \times 10^7$ cm^{-3} at 80 km.

Similar analysis was made for the flare of March 27, 1969 (1630–1700 UT). For this event the following values were obtained:

(a) $\alpha_i = 1 \times 10^{-6}$ cm^3/sec
  $\lambda_0 = 5$, $q_1 = 4$ cm^{-3} sec^{-1}
  $\lambda_0 = 10$, $q_1 = 25$ cm^{-3} sec^{-1}
(b) $\alpha_i = 1 \times 10^{-7}$ cm^3/sec
  $\lambda_0 = 5$, q_1 negative
  $\lambda_0 = 10$, $q_1 = 3.5$ cm^{-3} sec^{-1}.

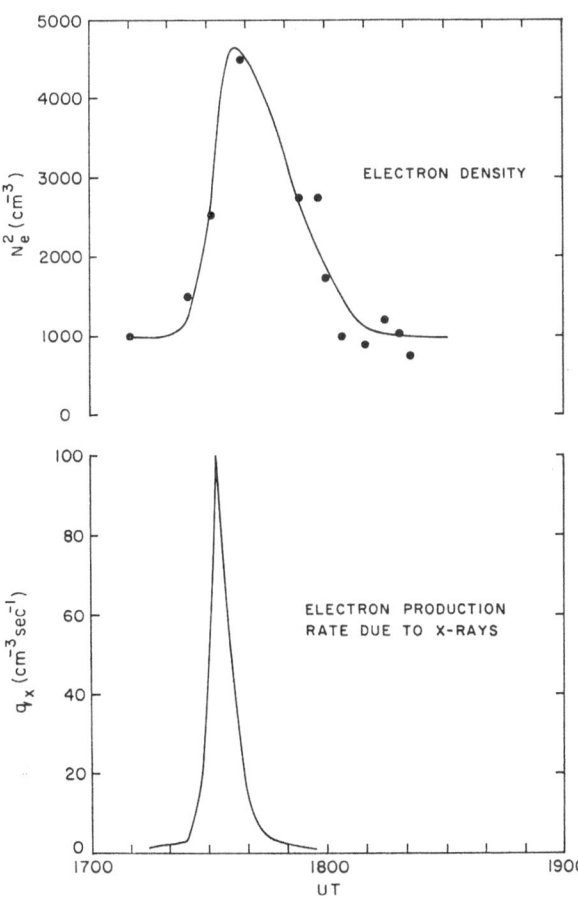

Fig. 11.   Estimated time variation of electron density for 80 km for the flare of October 21, 1968, along with the variations in electron production rate and effective loss rate.

A result of immediate interest is that, with variable λ, q_1 could be as large as 8 cm^{-3} sec^{-1}. This would correspond to an NO concentration of 2.4×10^7 cm^{-3}, whereas the constant loss rate analysis gives $n(NO) \simeq 3 \times 10^6$ cm^{-3}.

Similar analyses were made for other heights. Estimates of NO from the above considerations have resulted in the profile marked RFLKM which is shown along with the previous ionospheric estimates of Mitra, the rocket results of Barth (1966), Pearce (1969), Pontano (1970), and the new results of L. G. Meira (private communication). At 80 km there appears to be a general convergence to about 2×10^7 cm^{-3} and at 65 km the spread of value is reduced to an order of magnitude. The agreement between the ionospheric (flare) and rocket estimates of NO are now close enough to be considered satisfactory.

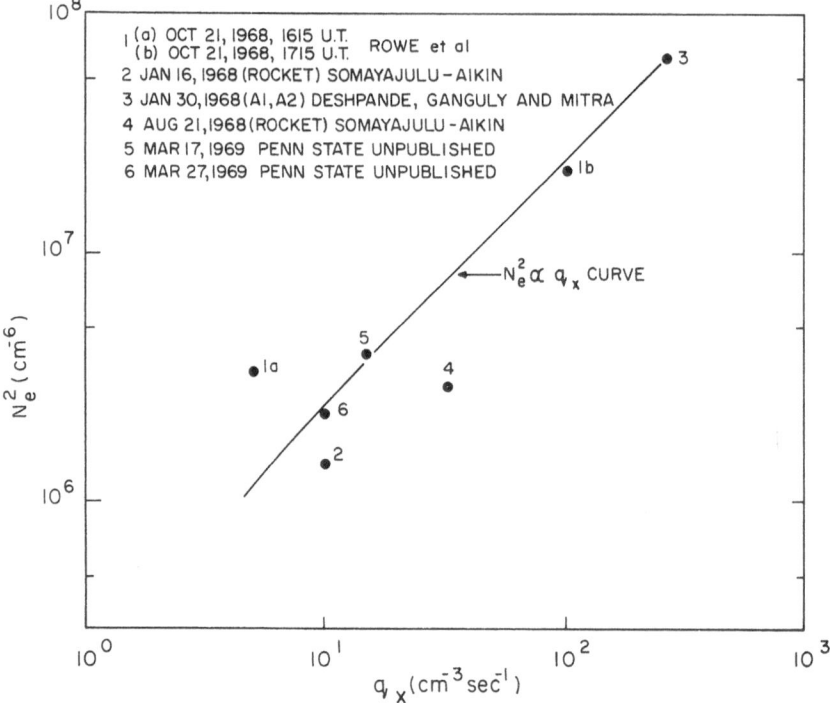

Fig. 12. N_e^2 during the peak of the flare plotted against electron production rate (q_x) due to X-rays for a number of flare events.

It is relevant in this connection to examine how N_e^2 varies with q_x for different flare events. Obviously if q_1 is large compared to q_x, this would be apparent in such a diagram. Figure 12 shows a plot for six flare events from Table I. The full lines gives the variation for the case where N_e^2 is directly proportional to q_x. In the two cases (Oct. 31, 1968 and Jan. 30, 1968) for which q_x was large, there is no doubt that such is indeed the case. The proportionality appears to extend down to $q_x = 10$–20 cm^{-3}. This would then confirm the conclusion given above.

3.3. LOSS RATE DURING FLARES

For the October 21, 1968 flare Rowe *et al.* find a large variation in the effective loss rate ψ which is shown in Figure 13. One notices that ψ decreases by a factor of more than 5 during the peak of the event.

The decrease in ψ would result in an overall decrease in the effective loss rate during a flare. Earlier estimates of the effective recombination coefficient from analysis of

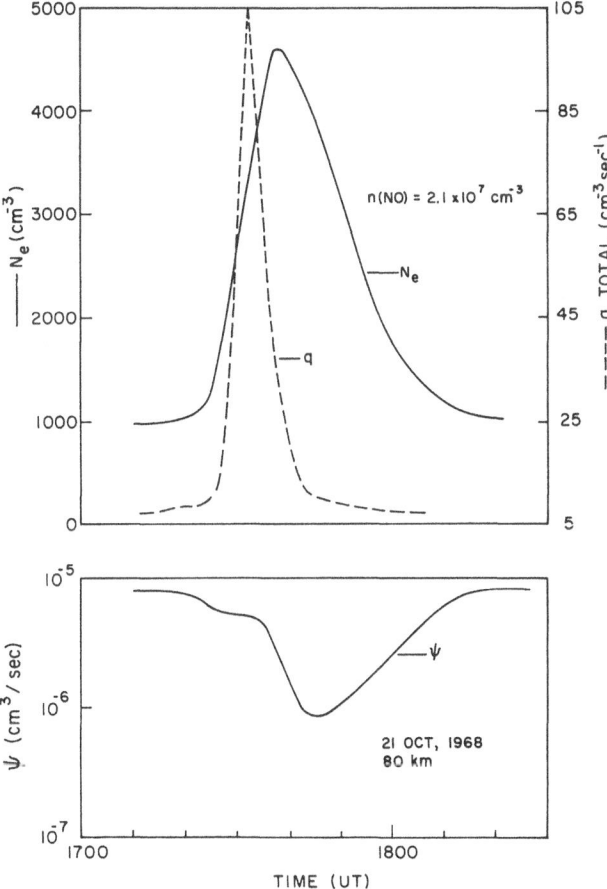

Fig. 13.   The total electron production rate $q(t)$ and the loss rate $\psi(t)$ during the flare of October 27, 1968, for 80 km. Note that a decrease in ψ occurs during the flare.

SPAs at LF/VLF, (Mitra, 1958; Mitra and Jones, 1954) and from long-wave field anomalies (S. N. Mitra, 1964; Entzian, 1964) are shown in Figure 14. Curiously, Mitra's analysis of SPA $\phi-t$ curves show a tendency for α_{eff} to decrease more sharply at about 70 km than the normal curve.

Deshpande and Mitra (1970) have recently determined the effective loss rates from

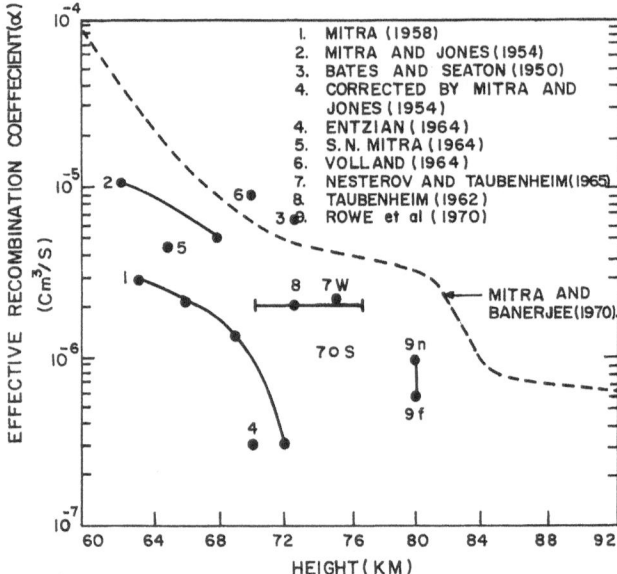

Fig. 14. Height distribution of the effective recombination coefficient during solar flare conditions.

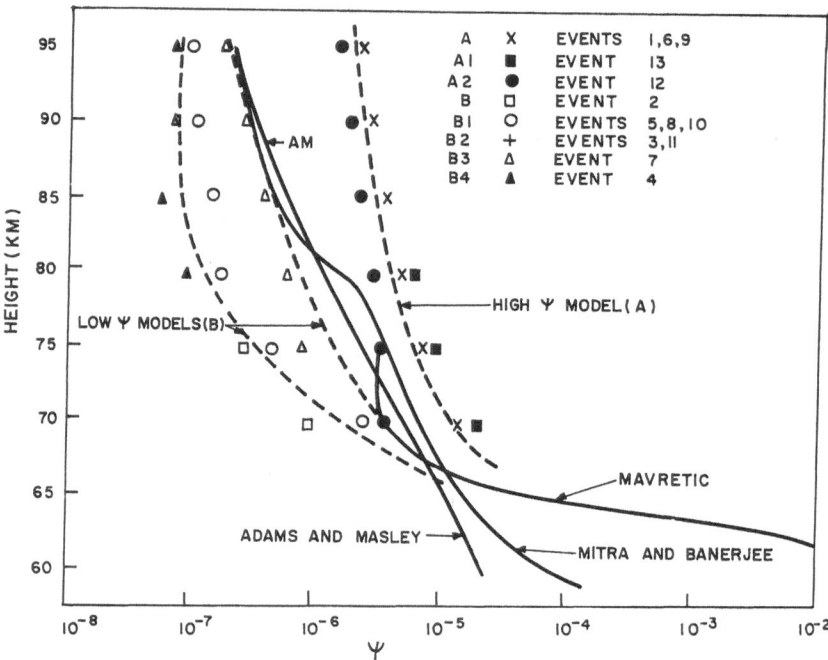

Fig. 15. Loss rates deduced from a comparative study of multifrequency SCNAs and satellite recorded X-rays. Note that the distributions of α during flare conditions can be quite different during different flares.

measurements of flaretime X-ray flux with detectors aboard SOLRAD 9 and corresponding SCNAs for about a dozen events. Different ψ models were tried out to produce the observed SCNA effect within $\pm 10\%$. Two assumptions were made:

(1) Firstly, it was assumed that ψ levels between 60 to 65 km is high, as would be given by Mavretic's λ model (1969).

(2) Secondly, ψ decreases continuously and smoothly with height. The rate of decrease may, however, be slow or rapid.

The ψ values derived in this way are marked A and B in Figure 15, and are seen to fall in two categories. In the first ψ has high values at all heights (models of type A); in the second (type B) ψ values are much lower and approximate either the model of Adams and Masley (1965) or of Mitra (1968). The nature of the flare radiation producing the type A or type B distributions has not been examined in detail. Further study could be very promising.

4. F Region During Flares

Since F-region ionization changes (only a few percent) are considerably smaller than in the D region, they are more difficult to investigate. The measurement techniques include: SFDs (at one or more frequencies), occasional cases of vertical incidence soundings at quick intervals, satellite radio beacon experiments with a geostationary satellite, and incoherent scatter 'soundings', so far only reported for the flares of May 21 and 23, 1967.

4.1. IONIZATION CHANGES

The incoherent-scatter results are given in Figure 16. These are obtained at Arecibo from simultaneous observations of the following: total power in the ion component of the backscattered spectrum as a function of height and time on 430 MHz; plasma-line echoes observed in 0.5 MHz steps from 3.5 to 9.5 MHz below the transmitted frequencies; phase-path length variations and absorption on two frequencies reflection in the F region. The results indicate (Thome and Wagner, 1967):

(1) Fairly large enhancements in the E region, decreasing with height in the F region, but with discernible enhancements up to 300 km.

(2) Delayed time of onset of the enhanced ionization at levels above the E layer, becoming most pronounced at F-layer heights.

(3) A negative fluctuation at F-region heights over the flare-affected region.

From ionograms taken at Kodaikanal at 15-minute intervals during the proton event of July 7, 1966, Bhattacharyya and Balakrishnan (1967) have derived changes of electron density at six levels in the 160–220 km range. These changes are shown in Figure 17. The normal day mean electron densities are shown by dotted curves. There was an increase in electron density at all the heights plotted, beginning at nearly the same time as the optical flare. There were two maxima; the authors attributed the first (occurring at about 0630 IST) to the solar XUV radiation, and the second (occurring immediately after 0700 UT) to collisional ionization by high-energy particles

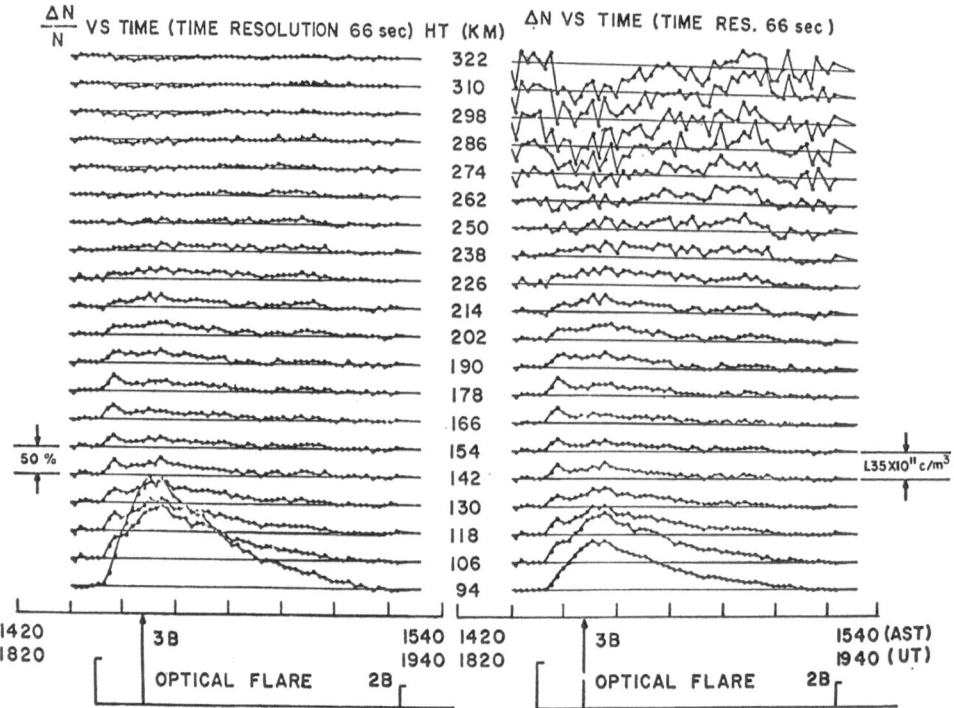

$\frac{\Delta N}{N}$ VS TIME (TIME RESOLUTION 66 sec) HT (KM) ΔN VS TIME (TIME RES. 66 sec)

Fig. 16.   Time profiles of enhanced electron densities obtained from backscatter power fluctuations at selected heights for the flare of May 21, 1967 (after Thome and Wagner, 1967).

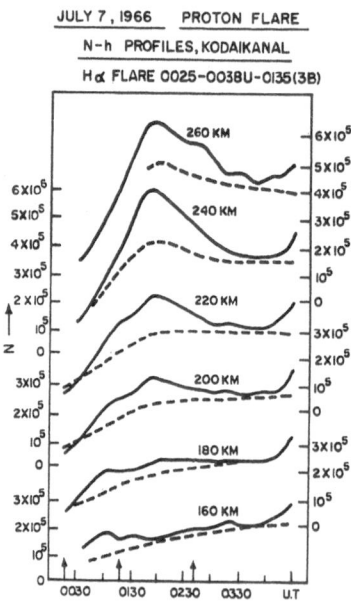

JULY 7, 1966   PROTON FLARE

N-h PROFILES, KODAIKANAL

Hα FLARE 0025–0038U–0135(3B)

Fig. 17.   *N-h* profiles deduced from ionograms taken at Kodaikanal during the solar proton flare of July 7, 1966 (after Bhattacharyya and Balakrishnan, 1967).

from the sun. We also note that there is a progressive delay in the time of maximum ionization as we go to higher levels.

Profile studies have also been made by Donnelly from observations of SFDs. An example is given in Figure 18. This was derived for the proton flare of August 28, 1966, from oblique incidence (Illinois-Boulder) SFD observations on 8.9 and 11.1 MHz and vertical incidence observations on 5.054, 4.000, 3.300 and 2.108 MHz. This was an unusually large SFD event.

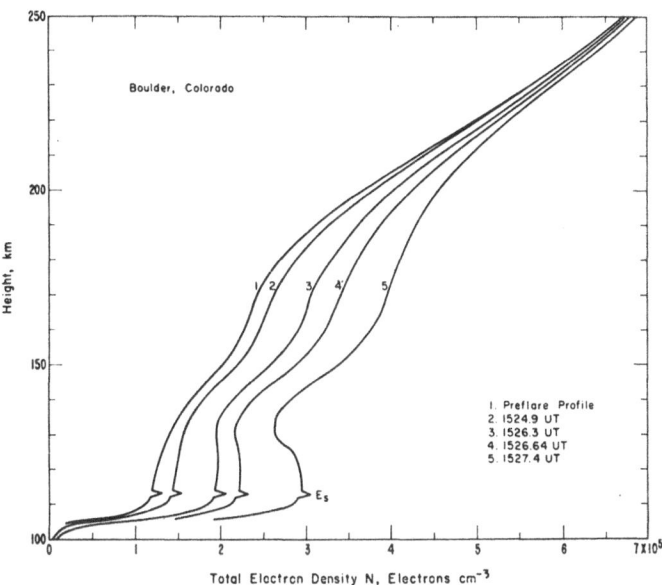

Fig. 18.   Boulder electron-density profiles during a flare deduced entirely from SFD data (after Donnelly, 1969).

It is remarkable that the profiles of Figures 17–18 are all roughly similar and ΔN values are comparable. In Table IV we give the ΔN values for selected heights for the different flare events discussed above.

Donnelly's profiles extend up to 300 km where the increase in ionization was estimated to be less than 1%.

For the May 1967 events probed by the Arecibo incoherent scatter equipment,

TABLE IV

Peak enhancement in electron density during different flare events

| Heights (km) | Incoherent Scatter | | Ionosonde | SFDs |
	May 21, 1967	May 23, 1967	July 7, 1966	August 28, 1966
160	0.35×10^5	0.6×10^5	0.8×10^5	$1.3 \ \times 10^5$
200	$0.35 \times 10^{5\,a)}$	0.7×10^5	$1 \ \times 10^5$	0.74×10^5

a) At 190 km.

Garriott *et al.* (1969) have reported at Stanford, San Diego, Flagstaff and Ely from the observations of the VHF telemetry signals received from the geostationary satellite ATS-1 at 137.35 MHz. Increases in electron content of about 2×10^{16} el/m^2 (about 5% of the total ionospheric content) were observed. The smaller value is expected because of the contributions of heights above 200 km where ionization enhancement is only a few percent.

4.2. ESTIMATES OF IONIZING RADIATION FROM F-REGION SIDs

The F-region SID effects are principally caused by flare radiation in the 100–1030 Å band but at E-region heights X-ray contribution from 10–100 Å band can be competitive. Since the time histories in the EUV region are at present available for only a few events, and these are known to differ from one event to another (Kreplin *et al.*, 1969), it is important to examine to what extent F-region SIDs can provide information on this radiation.

Detailed studies in this line are available from (a) ionogram analysis (Bhattacharyya and Balakrishnan, 1967), (b) SFD analysis (Donnelly, 1969) and (c) incoherent scatter *N-h* curves (Garriott *et al.*, 1969).

The ionogram analysis of Bhattacharyya and Balakrishnan refers to the event of July 7, 1966, given in Figure 17. They divided the spectrum into two bands: 44–100 Å and 200–900 Å, for which the production rates varied with height in the manner shown in Figure 19. Curves III and IV are their estimates, from the flare-time ionization profiles and assumed loss rates, for the actual production rates during the flare peak

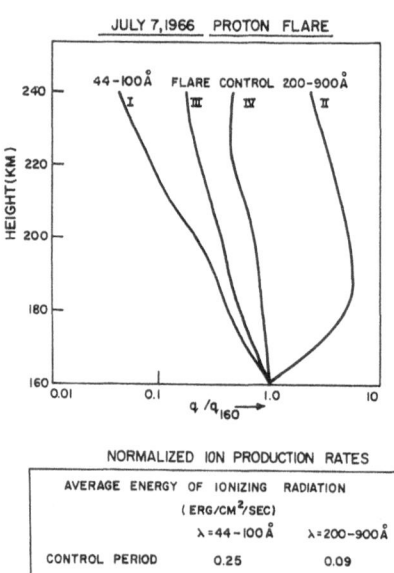

Fig. 19.   Height distributions of electron production rates due to X-rays (44–100 Å) and UV (200–900 Å) during the flare of July 7, 1966, compared to those during control conditions (after Bhattacharyya and Balakrishnan, 1967).

and during the control days. The production rates are normalized with respect to the 160 km value.

The prominent feature is that the q-profile is pulled towards the X-ray side during the flare.

The relative contributions of the X-rays and the EUV to F-region ionization have been examined in great detail by Donnelly on the basis of his very extensive SFD observations. SFDs are high time resolution broadband detectors of impulsive flare radiation in the 10–1030 Å range. They are insensitive to smooth enhancements of radiation with rise times greater than 5 min, because then the loss process nearly balances the production rate. Donnelly (1968) finds high energy X-rays ($\lambda < 0.6$ Å) and Lα enhancements do not contribute significantly to SFDs, 0.6–10 Å X-ray bursts are of minor importance to SFDs, 10–100 Å X-ray bursts do contribute significantly to SFDs (a point in agreement with Bhattacharyya and Balakrishnan), and flashes of radiation at wavelengths longer than 100 Å are required to explain most SFDs. In a more recent work (1969) he compares SFDs with OSO-3 EUV measurements from AFCRL and Explorer 30 X-ray measurements from NRL to further resolve the wavelength dependence of the radiation responsible for SFDs. The comparison was three-fold – a comparison of time dependence, intensity and transmission frequency dependence. It was observed that the EUV enhancements in He II 303.8 Å (see for example Figure 20), O v 639.7 Å, Ly 972.5 Å, C III 977.0 Å and Hα 1215.7 Å had the

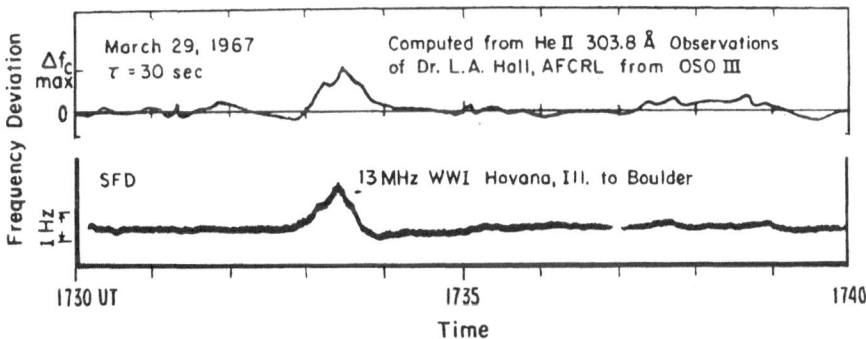

Fig. 20. Comparison of the time dependence of normalized frequency deviation computed from He II 303.8 Å flare measurements and the observed frequency deviation for UV flare of 1733 UT, March 29, 1967 (after Donnelly, 1969).

same time dependence as the net 1–1030 Å radiation responsible for SFDs, except that they decay faster than the SFD radiation. The coronal lines (Fe xv 284.1 Å, Fe xvi 335.3 Å, Si xii 499.3 Å, Mg x 625.3 Å, and Ne viii 770.4 Å), however, had a much slower time dependence than the SFD radiation. X rays in 0.5–3 Å and 0–8 Å band were slower during the decay stage and in the 8–20 band slower throughout the event than the SFD radiation. The largest contribution to the SFD was identified to be the hydrogen recombination continuum (14%). The largest contribution from the EUV lines studies was from C III 977.0 Å and was about half that from the hydrogen continuum. Donnelly concludes that the EUV recombination continuum and emission from the more abundant solar constituents H, He, O, C and N are the main cause of SFDs on paths reflected from the F region.

Another interesting spectrum building has been done by Garriott *et al.* (1969) from $N(h, t)$ curves obtained from the incoherent-scatter observations reported earlier. The solar radiation in the entire spectral region 1–1027 Å was divided into 5 wavelength groups with similar ionization and absorption cross sections as shown in Figure 21.

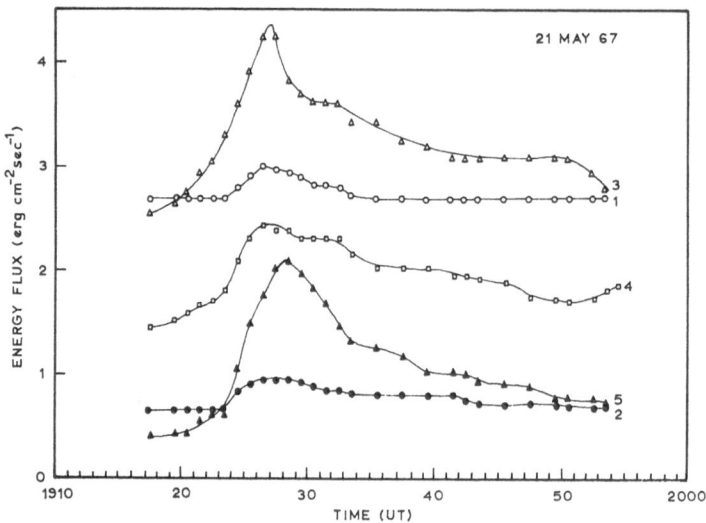

Fig. 21.   Time histories of the solar energy flux in the following 5 wavelength groups deduced from electron density changes given in Figure 18: 1. $\lambda\lambda$ 280–796 Å; 2. $\lambda\lambda$ 205–280 Å; 3. $\lambda\lambda$ 138–205 Å, 796–911 Å; 4. $\lambda\lambda$ 62–138 Å, 911–1027 Å; 5. $\lambda\lambda$ 1–60 Å (after Garriott *et al.*, 1969).

The groups are selected so that the heights of maximum electron production rate are separated by approximately one scale height in each case. The intensity in each group was allowed to vary in such a way that the computed ratio $R_{th} \left[= \dfrac{q(h, t)}{q(h, 0)} \right]$ matched the observed ratio, R_{exp} as closely as possible. It was observed that $R(h, t)$ increases at lower altitudes, indicating a hardening in the flare spectrum. Time histories of the derived flux in the five bands for the flare event of May 21, 1967 and corresponding to the ionization changes given in Figure 16 are shown in Figure 21. The time history of the total ionizing flux is also shown. The flux rose rapidly to a peak value of 4.8 erg cm^{-2} sec^{-1} at 1927 UT, about 8 min after the beginning of the flare effect, and then decayed much more slowly. Half-an-hour after the beginning of the event, the ionizing radiation was still above normal. A comparison of 1–60 Å curve with that of 1–12 Å band recorded by Explorer 33 show that they are essentially similar with the three-minute time resolution of the Geiger counter data, and would indicate that the analysis is reliable.

4.3. Loss rates during flares

Calculations of loss rates in the F region during flare events are scarce. Donnelly (1969) has determined α_{eff} using OSO-3 measurements of HeII 303.8 Å, Ov 629.7 Å, HLy 972.5 Å and CIII 977 Å along with Doppler frequency changes in SFDs. He obtained $\alpha_{eff} \simeq 2 \times 10^{-7}$ cm^3 sec^{-1} at 100 km and 1×10^{-7} at 160 km. The first is entirely consistent with current laboratory data; the second is somewhat larger than, but still comparable with laboratory data. No work exists to show that there has been any change in the loss rate or in the ion chemistry of the F region during a flare. Ion composition measurements would be valuable in this respect.

Acknowledgements

In the preparation of this paper much help was received from Mr. S. Ganguly and Miss Sudesh Kumari of this laboratory and from Mr. John Rowe of the Ionosphere Research Laboratory, Pennsylvania State University. The work has been supported in part by PL-480 Project 'Study of Flare Effects of Frequency Deviations and Phase Anomalies' (Agreement No. E-2-68(N)).

References

Adams, G. W. and Masley, A. J.: 1965, *J. Atmospheric Terrest. Phys.* **27**, 289.
Appleton, E. V. and Piggott, W. R.: 1954, *J. Atmospheric Terrest. Phys.* **5**, 141.
Bain, W. C. and May, B. R.: 1967, *Proc. IEEE* **114**, 1593.
Baker, D. C.: 1969, Sci. Rep. 334, Ionospheric Research Laboratory, Pennsylvania State University.
Barth, C. A.: 1966, *Ann. Geophys.* **22**, 198.
Bates, D. R. and Seaton, M. J.: 1950, *Proc. Phys. Soc.* **63**, 179.
Belrose, J. S.: 1969, Third Aeronomy Conference-Meteorological and Chemical Factors in D-region Aeronomy Rep. 32, University of Illinois, 375.
Belrose, J. S. and Cetiner, E.: 1962, *Nature, London* **195**, 688.
Bhattacharyya, J. C. and Balakrishnan, T. K.: 1967, *J. Atmospheric Terrest. Phys.* **29**, 1573.
Bracewell, R. N. and Straker, R. W.: 1949, *Monthly Notices Roy. Astron. Soc.* **109**, 28.
Burgess, B. and Jones, T. B.: 1967, *Radio Sci.* **2**, 619.
Deshpande, S. D. and Mitra, A. P.: In course of publication.
Deshpande, S. D., Ganguly, S., and Mitra, A. P.: In course of publication.
Deshpande, S. D., Subrahmanyan, C. V., and Mitra, A. P.: 1970, *Solar Flares,* I, 'Statistical Examination of X-ray Flare-SID Relationship, in press.
Donnelly, R. F.: 1968, *Solar Phys.* **5**, 123.
Donnelly, R. F.: 1969, *J. Geophys. Res.* **74**, 1873.
Ellison, M. A.: 1950, *Publ. Roy. Obs. Edinburgh* **1**, 53.
Entzian, G.: 1964, Vontragender Sommerschule, Kuhlungsborn/Heiligendaum.
Findlay, J. W.: 1951, *J. Atmospheric Terrest. Phys.* **1**, 367.
Gardner, F. F.: 1959, *Australian J. Phys.* **12**, 42.
Garriott, O. K., da Rosa, A. V., Davis, M. J., Wagner, L. S., and Thome, G. D.: 1969, *Solar Phys.* **8**, 226.
Houston, R. E., Ross, W. J., and Schmerling, E. R.: 1957, *J. Atmospheric. Terrest. Phys.* **10**, 136.
Kreplin, R. W., Horan, D. M., Chube, T. A., and Friedman, H.: 1969, *Solar Flares and Space Research*, North-Holland Publishing Co., Amsterdam, p. 121.
Mavretic, A.: 1969, Ionosphere Res. Laboratory, Pennsylvania State University, Sci. Rep. No. 334.
May, B. R.: 1966, *J. Atmospheric Terrest. Phys.* **28**, 533.

McIntosh, D. H.: 1951, *J. Atmospheric Terrest. Phys.* **1**, 315.

Mitra, A. P.: 1958, Ionosphere Res. Laboratory, Pennsylvania State Univ., Sci. Rep. No. 112.

Mitra, A. P.: 1960, Ionosphere Res. Laboratory, Pennsylvania State University, Sci. Rep. No. 142.

Mitra, A. P.: 1968, *J. Atmospheric Terrest. Phys.* **30**, 1065.

Mitra, A. P.: 1969, Third Aeronomy Conference-Meteorological and Chemical Factors in D-Region Aeronomy, Aeronomy Report No. 32, University of Illinois, 174.

Mitra, A. P. and Banerjee, P.: 1970, Paper presented at COSPAR Assembly, Leningrad, U.S.S.R.

Mitra, A. P. and Jones, R. E.: 1954, *J. Geophys. Res.* **59**, 39.

Mitra, A. P. and Sarada, K. A.: unpublished.

Mitra, A. P., Subrahmanyan, C. V., and Darabin Mirjana: 1964, *J. Atmospheric Terrest. Phys.* **26**, 1138.

Mitra, S. N.: 1964, *J. Atmospheric Terrest. Phys.* **26**, 375.

Nestorov, G. and Taubenheim, J.: 1965, *Izu. Geofiz. in-ta Bolgarsk An.* **7**.

Pearce, J. B.: 1969, *J. Geophys. Res.* **74**, 853.

Pontano, B. A.: 1970, Ionosphere Research Laboratory, Pennsylvania State University Scientific Report No. 347.

Rowe, J. H., Ferraro, A. H., Lee, H. S., Kreplin, R. W., and Mitra, A. P.: 1970, *J. Atmospheric Terrest. Phys.* **32**, 1609.

Sachdev, D. K.: 1958, *J. Sci. Indust. Res.*, July.

Sakurai, K.: 1968, *Geomagn. Geoelec.* **20**, 271.

Shain, C. A. and Mitra, A. P.: 1954, *J. Atmospheric Terrest. Phys.* **5**, 316.

Somayajulu, Y. V. and Aikin, C. A.: 1969, Third Aeronomy Conference-Meteorological and Chemical Factors in D-Region Aeronomy, Aeronomy Report No. 32, University of Illinois, 373.

Somayajulu, Y. V. and Aikin, C. A.: 1970, *Space Res.* **XI** (to be published).

Taubenheim, T.: 1962, *J. Atmospheric Terrest. Phys.* **24**, 1961.

Thome, G. and Wagner, L.: 1967, 'Solar Flare Effects in the E and F Regions of the Ionosphere', Paper presented at the Fall USNC-URSI meeting, Ann Arbor, Michigan.

Volland, H.: 1964, *J. Atmospheric. Terrest. Phys.* **26**, 695.

Weeks, K. and Stuart, R. D.: 1951, *Proc. Inst. Elec. Eng.* (London) **99**, 1.

BEHAVIOUR OF IONIZED COMPONENTS IN THE CHEMISTRY AND AERONOMY OF D AND E REGIONS

A. D. DANILOV

Hydrometeorological Service of the U.S.S.R., Moscow, U.S.S.R.

Abstract. The main problem of the ionospheric E region is that of the night source of ionization. Several sources are discussed: scattered Lα and Lβ emission, corpuscular fluxes. The latter were recorded in a number of rocket experiments, and their energy is enough for explanation of observed rate of recombination during the night. The behaviour of ion concentrations ratio NO^+/O_2^+ with height and time is connected with additional processes of transformation of O_2^+ ions to NO through ion-molecular reactions. The main problems of the D region physics are: whether, or not are true measured high concentrations of positive ions and nitric oxide. Estimates, based on measured values of $[e]$ and α' give us low values of λ in contradiction with probe measurements. The modern scheme of negative-ion transformations can explain some experimental data if an additional source of conversion O_3^- ions to NO_2^- is suggested.

1. Introduction

During the last years so many papers have appeared concerning the behaviour of ionized particles in the lower ionosphere (D and E regions) that it is impossible to describe or even to mention all of them in one report. It is impossible also to discuss in detail all the problems of ionospheric physics below 130–140 km, so we shall try to give only a general description of the main problems, emphasizing problems where contradictions between theoretical points of view and experimental data are seen most clearly.

2. E Region

The most interesting question of ionospheric physics at altitudes 100–140 km is that of the night source of ionization in the E region. For a long time it was assumed that the effective recombination coefficient in this region was of the order of 10^{-8} cm^3 sec^{-1} (Bates, 1956; Ratcliffe and Weekes, 1960; Mitra, 1959). This value was arrived at on the basis of the suggestion that the observed slow decrease of electron concentration during the night is due only to recombination processes without any additional formation of electrons after sunset. But at the same time the observations show that during the day the effective recombination coefficient in the E region is much larger: of the order of 10^{-7} cm^3 sec^{-1} and at some periods even higher (Serafimov, 1970; Ivanov-Kholodny and Nicolsky, 1969). This is in good agreement with modern photochemical ideas, since the value $\alpha' = 10^{-7}$ cm^3 sec^{-1} is obtained (Ivanov-Kholodny and Nicolsky, 1969; Mitra, 1963) on the basis of the well-known process of dissociative recombination. But a reduction of α' in the E region from day to night down to the value 10^{-8} cm^3 sec^{-1} appears to be impossible from the point of view of photochemical theory. Firstly, because of the increase at night in the relative proportion of NO^+ ions, which have a higher value of dissociative recombination coefficient than

Dyer (ed.), Solar-Terrestrial Physics/1970: Part IV, 27–40. All Rights Reserved.
Copyright © 1972 by D. Reidel Publishing Company.

O_2^+ ions; and secondly, because of the decrease in electron temperature T_e, giving an increase after sunset in the value of α'.

The behaviour of $[e]$ in the E region and above after sunset is shown in Figure 1, taken from a review of rocket measurements of electron concentration (Ivanov-Kholodny and Kazachevskaya, 1966). One can see from this picture that the observed decrease of $[e]$ at night corresponds to an even lower value ($\alpha'_E \approx 10^{-8}$ cm^3 sec^{-1}) than that corresponding to the decrease of electron concentration at heights 125–150 km (where $\alpha' \approx (3-10) \, 10^{-8}$ cm^3 sec^{-1}). Such a behaviour of α' with altitude also seems to

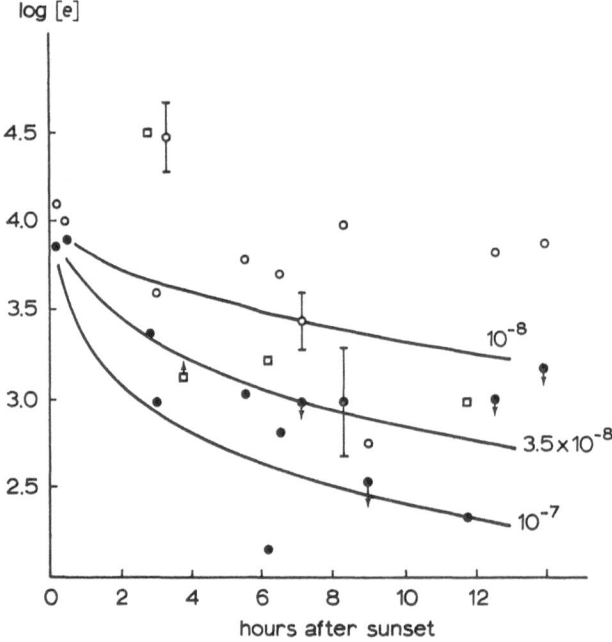

Fig. 1.   Behaviour of electron concentrations after sunset at different heights according to the measurements (□ – 110 km, ○ – 120 km, ● – 125–160 km) and only in case of recombination with various α' values (curves) (Ivanov-Kholodny and Kazachevskaya, 1966).

contradict modern ideas on recombination processes in the ionosphere. It is simple to avoid these contradictions, however, if one supposes that the electron concentration is supported by an additional source of ionization at night. In this case, the observed slow decrease of electron concentration after sunset reflects not a low value of α'_E at night, but the decrease in intensity of the additional source in comparison with the solar ultraviolet, which acts during the day. The necessity of an additional source of ionization in the E region can be clearly seen from Figure 2, due to Ivanov-Kholodny and Korsunova (1970). It shows the variation at altitudes 100–125 km of the values of q/α' and $[e]^2$, should be equal under conditions of photochemical equilibrium, for different solar zenith angles for low solar activity. One can see from the figure that for solar zenith angles $\chi = 70°$, the values of q/α' and $[e]^2$ agree well, showing that, at that time, ionization by solar ultraviolet is completely responsible for the observed $[e]$ values. But for $\chi = 85°$ there is a pronounced difference between the curves for q/α' and $[e]^2$ and the difference is greater for $\chi = 90°$. Thus, it appears that at the time of sunset the observed electron concentration cannot be explained with high values of α'_E by solar radiation alone; an additional source of ionization is required. Another point in favour of the existence of a night source of ionization is the observed change of ion composition from day to night. Because α^* is higher for NO^+ than for O_2^+, evidently during the night decrease of $[e]$ due to recombination, the relative fraction

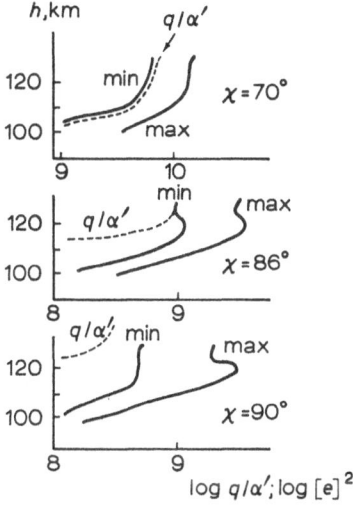

Fig. 2. A comparison between q/α' and $[e]^2$ in the E region for different $\chi \odot$ according to Ivanov-Kholodny and Korsunova (1970).

of NO^+ ions has to decrease and become smaller at night than in the day. However, according to experimental data (Istomin, 1963; Holmes et al., 1965; Narcisi, 1967) the reverse situation applies – the night values of $[NO^+]/[O_2^+]$ are higher than the day values.

The question of the nature of this additional source is at present the subject of intense discussion (Ivanov-Kholodny and Korsunova, 1970; Yonezawa, 1968; Swider and Keneshea, 1968; Swider, 1965; Ivanov-Kholodny, 1969). In some papers (see for example Swider and Keneshea, 1968; Swider, 1965; Tohmatsu and Ogawa, 1966) it is suggested that the source of the twilight and night E region is ionization of NO molecules by scattered Lα flux. The role of this source depends upon the concentration of nitric oxide at altitudes of 90–120 km. But Swider and Keneshea (1968) find that even with high values of $[NO]$ measured by Barth (1966) this process can be important only at altitudes 90–100 km; in the higher E region there must exist another source of night ionization, for example, ionization of O_2 by scattered Lβ flux. The main difficulty for computation of the role of both these possible mechanisms lies in the fact that the

intensity of the scattered radiation and penetration of the latter into the atmosphere
are poorly known.

It is possible, however, to argue against a leading role for these two sources in the
ionization of the night E region. It is known from experiments that $[e]$ values in the
E region vary considerably from night to night, while it is difficult to accept corresponding (proportional to $[e]^2$) variations in the intensity of scattered solar radiation.
Also, the fact of the increase of $[e]$ with latitude and with increasing geomagnetic
activity (Kazachevskaya and Ivanov-Kholodny, 1970; Korsunova, 1966) is hard to
explain on the basis of ionization of the E region by scattered radiation in the Lα
and Lβ lines.

The theory (Ivanov-Kholodny and Korsunova, 1970; Ivanov-Kholodny, 1965, 1969;
Kazachevskaya and Ivanov-Kholodny, 1970) of ionization of E region at twilight and
at night by a flux of soft electrons seems to be much more attractive. Such fluxes
of electrons with energy $E \geqslant 1-5$ keV were recorded in numerous experiments of
Ivanov-Kholodny and coworkers (Ivanov-Kholodny, 1969; Kazachevskaya and
Ivanov-Kholodny, 1970; Antonova et al., 1970; Antonova and Kazachevskaya, 1970).
On the basis of these experiments intensities of corpuscular fluxes were given as
shown in Table I.

TABLE I

	Measurements (erg cm^{-2} sec^{-1})	Ionospheric estimates (erg cm^{-2} sec^{-1})
Day	0.3–1	0.4
$\chi = 85-90°$	0.2	1.5×10^{-2}
Night	6×10^{-3}	$(0.3-20) \times 10^{-4}$

In this table are presented also the results of computation of energy of the ionizing
corpuscular flux on the basis of ionospheric data $(q = \alpha'[e]^2)$. One can see from
the table that the measured electron flux is enough for providing the energy, derived
from the recombination rate. Some indirect arguments can be given in favour of
corpuscular fluxes as a source of ionization in the E region. For example, corpuscular ionization can explain the strong variability of ionization from night to night
in the E region, because the above-mentioned experimental data show that the
intensity of corpuscular fluxes undergoes significant variation. It is possible also to
understand the increase of $[e]$ with latitude and with geomagnetic activity because
in both cases the importance of corpuscular ionization must increase.

Arguments in favour of the role of observed electron fluxes in E-region ionization
have been given by Kazachevskaya and Ivanov-Kholodny (1970). They found an
increase in E-region critical frequency simultaneously with an increase in corpuscular
flux.

It is necessary to mention the attempt of Yonezawa (1968) to solve the problem
without an additional source of ionization. He supposed the effective recombination

coefficient to be low ($\approx 10^{-8}$ cm^3 sec^{-1}) because the molecular ions were in an excited state and so had a low coefficient of dissociative recombination. However, it seems certain that a high coefficient ($\alpha'_E \approx 10^{-7}$ cm^3 sec^{-1}) exists in the E region and therefore the conclusion about an additional source of ionization at night is inevitable.

The photochemistry of the positive ions in the E region below 120 km does not differ from that in the region 120–200 km with one exception. At altitudes $h \geqslant 140$ km the ratio $[NO^+]/[O_2^+]$ changes only slightly with height and time (Danilov and Ivanov-Kholodny, 1966) while for $h < 130$–140 km an increase of this ratio at lower heights and during the night is observed. This behaviour of $[NO^+]/[O_2^+]$ ratio is explained by two reasons. Firstly in the E region there are two important ion-molecular reactions with the minor constituents N and NO

$$O_2^+ + N \rightarrow NO^+ + O \tag{1}$$

$$O_2^+ + NO \rightarrow NO^+ + O_2 \tag{2}$$

which lead to transformation of O_2^+ to NO^+ and consequently to the growth of $[NO^+]/[O_2^+]$ (Danilov, 1969; Ivanov-Kholodny and Korsunova, 1969). From day to night this ratio which is inversely proportional to $[e]$, has to increase because of decrease $[e]$. Secondly at altitudes below 130–140 km ionization of O_2 by radiation in the region 911–1027 Å plays an essential role. This radiation penetrates deeper into the atmosphere, than the main ionizing radiation, and causes an excess of O_2^+ ions. After sunset this additional formation of O_2^+ ceases and the ratio $[NO^+]/[O_2^+]$ in the E region has to increase. The observed increase of this ratio with solar activity may be explained in the same way, since the intensity of radiation in the region 911–1027 Å increases by less than a factor of 1.5 between minimum and maximum solar activity, and the intensity of the main ionizing radiation < 900 Å increases by a factor 4–5.

On the basis of the data on variations of $[NO^+]/[O_2^+]$ one can obtain some idea of the change in value of α' between day and night in the E region. If the electron temperature T_e at these heights during the daytime exceeds the night value by about 1.5–2 times, then, together with the increase at night in the fraction of NO^+ ions (with higher α^* coefficient), it would lead to an increase in α' from day to night by a factor 2–2.5 (Ivanov-Kholodny and Korsunova, 1969). This behaviour of α', as was mentioned above, contradicts observations showing that α' decreases after sunset. It seems possible, however that a correct allowance for night sources of ionization may eliminate these contradictions and bring about agreement between photochemical ideas and experimental results.

3. D Region

It has become usual to say that the physics of the D region is much less well known to us than the physics of the ionospheric regions above it. It seems that the reason is twofold: (1) measurements of charged-particle concentrations in the lower ionosphere are much more complicated than the same measurements above 100 km; and (2) an essential role is played in the physics of the lower ionosphere by minor-constituents such as NO, NO_2, O, H_2O and so on, concentrations of which are poorly known.

The first consideration leads to the situation that up to now we have no reliable sets of data on the main charged particles: $[e]$, $[X^+]$, $[X^-]$ which would (as was done, say, in the 100–200 km region) allow us to construct distinct systems for ionization and recombination reactions. The second consideration makes it very difficult to construct the true picture of changing positive- and negative-ion composition. As a result of the presence in the D region of H_2O molecules, positive ions transform into complicated hydrate clusters, the photochemistry of which is just beginning to be developed. For negative ions, the presence of NO, NO_2, CO_2 and O leads to the compound system of ion reactions with such ions as NO_3^-, CO_3^-, NO_2^- for which very important parameters (such as the rates of mutual neutralization or photo-detachment) are not known. The two difficulties described lead to the consequence that even such an important parameter of the lower ionosphere as λ (the ratio $[X^-]/[e]$) is essentially unknown.

There are three possible ways to estimate the value of λ. The first is to construct the system of reactions for the negative ions, and to find λ from photochemistry (see, for example Moler, 1960; Nicolet and Aikin, 1960; Danilov, 1970; Aikin, 1967). However it has become obvious that the simple scheme with only one ion (O_2^-) does not reflect the real conditions (LeLevier and Branscomb, 1968). More complicated schemes have not yet been developed well enough to give quantitative conclusions about λ.

The second way consists of using measured values of $[e]$ and $[X^+]$ calculating values of λ from the equation of quasi-neutrality of charge:

$$[X^+] = [e] + [X^-] = [e](1 + \lambda). \tag{3}$$

However the reliability of measurements of $[e]$ and $[X^+]$ in the lower ionosphere is different. For $[e]$ we have many measurements by various methods and it seems that mean $[e]$ values are known within a factor of 2 (Ivanov-Kholodny, 1964; Bowhill, 1969). But measurements of the total concentration $[X^+]$ of positively charged parti-cles are carried out mainly by direct probe methods, the reliability of which is dubious. These measurements as a rule give very high values: $[X^+] \approx (1-10) \, 10^3 \, \text{cm}^{-3}$ in the whole D region. This leads, for relatively low $[e]$ values, to high values for λ. So if one takes $[e]$ and $[X^+]$ from a recent paper by Bowhill (1969), one calculates, for day conditions $\lambda = 9$ at 80 km and $\lambda = 15$ at 75 km. The night values of λ have to be correspondingly higher.

TABLE II

Parameters of the D region from Bowhill (1969)

h, km	60	70	75	80
$[e]$	36	115	240	600
$[X^+]$	1300	2100	3400	6000
λ	30	20	15	9

The third and final way to determine λ is the following (Ivanov-Kholodny, 1967). It seems that the two parameters $[e]$ and α' are known best of all in the D region. Using them and equilibrium conditions for the electron concentration

$$\frac{d[e]}{dt} = \frac{q}{1+\lambda} - \alpha'[e]^2 = 0 \tag{4}$$

one can solve for λ for altitudes 50–60 km, where all the ionization is produced by cosmic rays and all three values q, α' and $[e]$ are so known. On the basis of the calculated values of λ it is possible from $\alpha' = \alpha^* + \lambda \alpha_{mn}$ to find α_{mn} (the coefficient of mutual neutralization of X^+ and X^-) and to use this coefficient for finding λ at higher altitudes from the same expression. The assumed values for $[e]$, α' and q and the computed values of λ are given in Table III (according to Ivanov-Kholodny, 1967).

TABLE III

Parameters of the D region

h, km	α'	$[e]$	q	λ
50	2.5×10^{-5}	20	0.35	35
60	8×10^{-6}	50	0.1	4
70	2.1×10^{-6}	160	–	1.1
80	4.7×10^{-7}	800	–	0.1

As can be seen from the table, the method described leads to considerably lower values of λ than measurements of X^+ with probes (see Table II). The weak points of the method are the following: (1) for $h > 60$ km a value of α_{mn} is used which was obtained for heights 50–60 km, and (2) λ values for 70 and 80 km are found from expression (4) on the basis of a difference between values α' and α^* which may be small.

However, it is impossible to use the method above described for calculating values of λ at $h > 60$ km directly from the $[e]$ and α' values because of uncertainties in q at these altitudes. It seems that the main question is now whether the high concentrations of NO measured by Barth (1966), Pearce (1969) and by Pontano and Hale (1970) are true or not. Using an average distribution of $[NO]$ (Danilov and Tresvatsky, 1970) on the basis of these experiments one calculates very high values of q (9 cm^{-3} sec^{-1} at 70 km and 50 cm^{-3} sec^{-1} at 80 km). On this basis, the comparison with $[e]$ values (see Table III) leads through Equation (4) to unrealistically high values of λ (140 at 70 km and 170 at 80 km), even higher than values, obtained from $[X^+]$ measurements. It seems that high concentrations of NO are also difficult to reconcile with other aspects of photochemistry: with systems of neutral reactions without too high a rate constant for the reaction $N + O_2$ ($^1\Delta g$) (Danilov and Tresvatsky, 1970; Hunten and McElroy, 1969) or strong production of $N(^2D)$ and with the system of reactions for NO^+ formation and disappearance (Mitra, 1969b).

If the above-mentioned measurements of NO in the D region are erroneous and the real concentrations of nitric oxide are several orders lower, than it is useful to discuss other possible mechanisms of ionization at altitudes 65–85 km.

The necessity for such a mechanism can be seen from Figure 3. The figure shows rates of ionization both by cosmic rays and solar X-rays, and $\alpha'[e]^2$ values for comparison (Ivanov-Kholodny and Nicolsky, 1969). Evidently some additional source of ionization is needed in the region 65–85 km. The following processes have been discussed in the literature: (a) ionization by flux of corpuscules with $E > 30$–50 keV (Ivanov-Kholodny and Nicolsky, 1969; Tulinov, 1967; Radicella, 1968), (b) ionization of excited molecules, $O_2(^1\Delta_g)$, by radiation in the range 1027–1118 (Hunten and McElroy, 1968; Vlasov, 1970); and (c) ionization of vibrationally excited O_2 molecules by $L\alpha$ radiation (Inn, 1961).

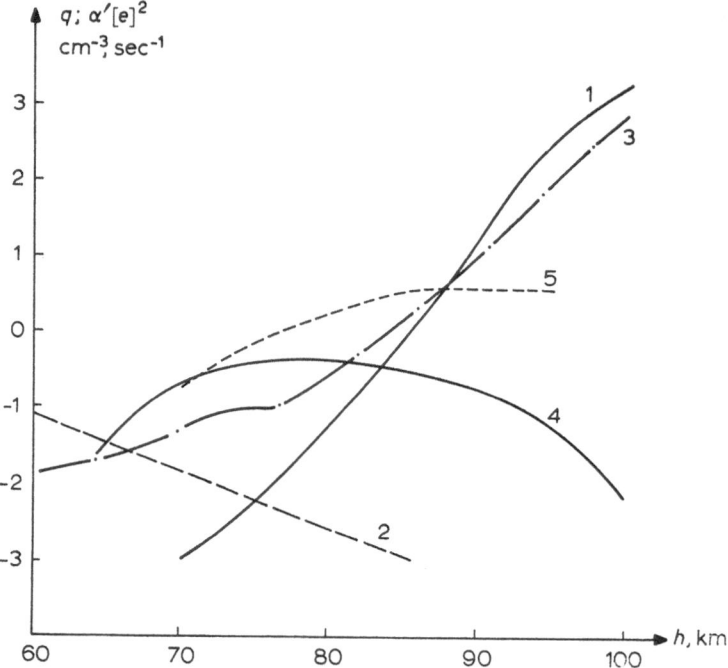

Fig. 3.   A comparison between q and $\alpha'[e]^2$ in the D region:
1 – the rate of photoionization by solar X-rays;
2 – the rate of ionization by cosmic rays;
3 – $\alpha'[e]^2$;
4 – the rate of photoionization of O_2^* by radiation; in the range 1027–1118 Å and by $L\alpha$;
5 – the rate of photoionization of $O_2(^1\Delta g)$ by radiation with $\lambda = 1027$–1118 Å;
Curves 1–3 are from Ivanov-Kholodny and Nicolsky (1969), curves 4 and 5 from Vlasov (1970).

It seems that the main sources of ionization at these heights during the day are processes (b) and (c), while corpuscular ionization may play a significant role in the night D region (Radicella, 1968). The rates of ionization for the processes:

$$O_2(^1\Delta_g) + h\nu(1027\text{–}1118 \text{ Å}) \rightarrow O_2^+ + e \qquad (5)$$

and

$$O_2^* + h\nu \rightarrow O_2^+ + e \qquad (6)$$

are given in Figure 3 according to Vlasov (1970). It can be seen, that according to computations ionization of excited O_2 molecules is able to provide the observed ionization at altitudes of 65–85 km. But the reliability of the $q(O_2^*)$ and $q(O_2^{\#})$ values is low, so it is possible only to conclude that the observed values $\alpha'[e]^2$ can be explained by these mechanisms, but not to compute values of λ by the method described above.

Let us now consider modern ideas about photochemistry of negative ions in the D region (Radicella, 1968; Fehsenfeld *et al.*, 1967; LeLevier and Branscomb, 1968; Mitra, 1969a; Adams and Megill, 1967).

The scheme of transformation of negative ions is presented in Table 4 and Figure 4. It is constructed on the basis of laboratory data reviews (Ferguson *et al.*, 1967;

TABLE IV

N	Reaction	Rate constant
1	$X + h\nu \rightarrow X^+ + e$	q
2	$O_2 + O_2 + e \rightarrow O_2^- + O_2$	1.5×10^{-30}
3	$O_2^- + O_3 \rightarrow O_3^- + O_2$	3.5×10^{-10}
4	$O_2^- + O_2(^1\Delta g) \rightarrow O_2 + O_2 + e$	2×10^{-10}
5	$O_2^- + O \rightarrow O_3 + e$	2.5×10^{-10}
6	$O_2^- + NO_2 \rightarrow NO_2^- + O_2$	8×10^{-10}
7	$O_3^- + NO \rightarrow NO_3^- + O$	1×10^{-11}
8	$O_3^- + CO_2 \rightarrow CO_3^- + O_2$	4×10^{-11}
9	$CO_3^- + O \rightarrow O_2^- + CO_2$	8×10^{-11}
10	$NO_2^- + O_3 \rightarrow NO_3^- + O_2$	1.8×10^{-11}
11	$NO^- + O_2 \rightarrow O_2^- + NO$	9×10^{-10}
12	$CO_3^- + NO_2 \rightarrow NO_3^- + CO_2$	9×10^{-11}
13	$NO_2^- + X^+ \rightarrow NO_2 + X$	?
14	$NO_3^- + X^+ \rightarrow NO_3 + X$	?

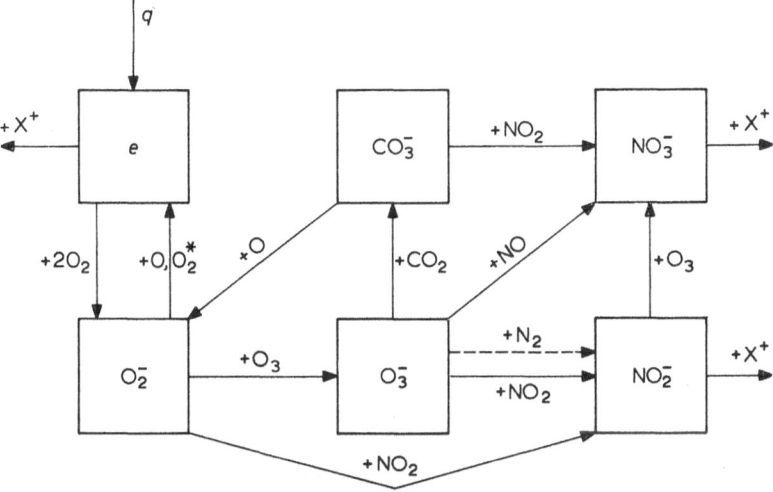

Fig. 4.   A scheme of negative ion transformations in the D region.

Phelps, 1969; Ferguson, 1969; Schoen, 1969). Before starting the analysis of the
behaviour of different negative ions, it is necessary to emphasize that all the conclusions on negative ion species concentrations X_i^- are strongly dependent on the
assumed concentrations of the minor constituents NO, NO_2, CO_2, O_3 and O. We have
used the model of neutral constituents, given in Table V.

TABLE V

h km	$[O_3]$ cm^{-3}	$[O]$ cm^{-3}	$[NO]$ cm^{-3}	$[O_2(^1\Delta_g)]$ cm^{-3}	$[NO_2]$ cm^{-3}
50	4.8×10^{10}	7×10^9	7×10^8	5×10^{10}	10^6
60	3.4×10^9	1.5×10^{10}	3.6×10^8	1.5×10^{10}	3×10^4
70	4.3×10^8	5×10^{10}	2.0×10^8	7×10^9	$< 10^4$
80	1.0×10^8	5×10^{10}	1.3×10^8	5×10^9	–
90	2.5×10^7	3×10^{11}	4.0×10^7	2×10^9	–

Analysis of rates of disappearance of O_3 ions show that, in the whole 50–90 km
region, the disappearance of these ions is due to reaction with CO_2 so:

$$[O_2^-] [O_3] \alpha_3 = [O_3^-] [CO_2] \alpha_8. \tag{7}$$

In the same way it is easy to show that the destruction of CO_3^- is due only to its
reaction with atomic oxygen, and so:

$$[O_3^-] [CO_2] \alpha_8 = [CO_3^-] [O] \alpha_9. \tag{8}$$

In that case the disappearance of O_2^- in reaction with O_3 is compensated by the
formation of O_2^- in reaction 9 in Table IV and corresponding terms may be excluded
from the balance equation for O_2^-:

$$[O_2]^2 [e] \alpha_2 + [CO_3^-] [O] \alpha_9 = [O_2^-] [O_3] \alpha_3 + [O_2^-] ([O] \alpha_5 + [O_2^*] \alpha_4) \tag{9}$$

which gives the following expression for $[O_2^-]$:

$$[O_2^-] = \frac{[O_2]^2 [e] \alpha_2}{[O] \alpha_5 + [O_2^*] \alpha_4}. \tag{10}$$

It follows from the formulas (7), (8) and (10) that within this scheme of processes
the concentrations of O_2^-, O_3^- and CO_3^- do not depend on ionization and recombination processes and are strongly connected with electron concentration. An example
of the calculation of O_2^-, O_3^- and CO_3^- ion concentrations on the basis of the values
of $[e]$ given in Table III, is presented in Table VI. As can be seen from this table,
above 60 km the sum of the concentrations of these ions is small (50 cm^{-3} at 60 km
and 5 cm^{-3} at 70 km).

The main difficulty of this scheme is as follows. The system $[e] \to O_2^- \to CO_3^- \to O_2^- \to [e]$
is closed, and thus cannot take part in the removal of electrons. The transformation
of these ions to NO_2^- and NO_3^- ions is slow (much less than q), so the whole disap-

TABLE VI

h km	$[O^-_2]$ cm^{-3}	$[O^-_3]$ cm^{-3}	$[CO^-_3]$ cm^{-3}	The rate of formation of NO$^-_3$ and NO$^-_2$
50	6.3×10^1	3.6	1.8×10^3	2.5×10^{-1}
60	2.5×10^1	3.6×10^{-1}	2.7×10^1	2.0×10^{-3}
70	3.4	2.0×10^{-2}	1.3×10^{-1}	4.2×10^{-5}
80	7.1×10^{-1}	4.8×10^{-3}	6.3×10^{-3}	6.2×10^{-6}
90	2.1×10^{-2}	2.2×10^{-4}	8.3×10^{-6}	–

pearance of electrons in that case has to be provided by dissociative recombination. But it leads (with $\alpha^* \leqslant 10^{-6}$ cm^3 sec^{-1}) to $[X^+] \geqslant 2500$ cm^{-3} at 60 km. Since in our computations $[e] + [O^-_2] + [CO^-_3] + [O^-_3]$ is small ($< 10^2$ cm^{-3}) then the concentrations of NO$^-_2$ and NO$^-_3$ must be about equal to $[X^+]$. It is then easy to estimate the rate of mutual neutralization which is necessary to support such high values of $[NO^-_2]$ or $[NO^-_3]$, and α_{mn} appears to be 4×10^{-10} cm^3 sec^{-1}. This value seems to be very small. But with more realistic values of α_{mn} ($10^{-6} - 10^{-8}$) it is impossible to accept (within the scheme described) high concentrations of NO$^-_2$ and NO$^-_3$, or in other words, to arrive at consistent values for q $[e]$ and α'.

It seems that there must be an effective additional way of converting O^-_2, O^-_3, or CO^-_3 ions into NO$^-_2$ or NO$^-_3$ ions (Radicella, 1968; LeLevier and Branscomb, 1968). It is quite probable the reaction:

$$O^-_3 + N_2 \rightarrow NO^-_2 + NO \tag{11}$$

suggested by Whitten and Poppoff (1964, 1965) and later by Radicella (1968) may be the required mechanism. The rate of the reaction has not been measured, but it is enough to have α_{14} of the order of 10^{-16} cm^3 sec^{-1} (this value is far below the sensitivity of laboratory measurements of negative ion-molecule reactions) to make the new scheme agree with the experimental data. The inclusion of reaction (11) does not disturb the cycle $e \rightarrow O^-_2 \rightarrow CO^-_3 \rightarrow O^-_2 \rightarrow e$, because the reaction with CO_2 remains still the main process of O^-_3 destruction. But now it is possible to achieve the proper $[X^-]$ with more realistic values of α_{mn}, of the order of $10^{-6} - 10^{-8}$ cm^3 sec^{-1}. For example assuming $\alpha_{14} = 6 \times 10^{-17}$ cm^3 sec^{-1} (to make the rate of formation of NO$^-_2$ agree with q at 60 km) we derive $\alpha_{mn} = 7 \times 10^{-8}$ cm^3 sec^{-1} if $[X^+] = 1300$ cm^{-3} (see Table II) and $\alpha_{mn} = 5 \times 10^{-7}$ cm^3 sec^{-1} if $[X^+]$ is less (500 cm^{-3}). However, such values of α_{mn} seem to be reasonable for mutual neutralization of relatively simple positive ions, such as NO$^+$ or O$^+_2$, but we know nothing about the values for complicated clustered ions such as H_3O^+, $H_5O^+_2$, $H_7O^+_3$, etc.

The photochemistry of cluster ions is now in the stage of development. The results of laboratory measurements allowed (Ferguson, 1969; Ferguson and Fehsenfeld, 1969; Fehsenfeld and Ferguson, 1969) to construct a scheme of chain reactions, which forms clustered ions in the D region. This scheme is able principally to explain the presence of ions with masses 19$^+$, 37$^+$ and 55$^+$; and the fact that ions with mass 37$^+$

predominate. At a height of about 80 km, Ferguson succeeded in achieving quantitative agreement with Narcisi's experimental data, but for lower heights the agreement was poor. The sharp decrease of cluster ion concentration above 82 km is also not explained.

The essential peculiarity of this scheme is that the formation of clusters proceeds better from O_2^+ than from NO^+ ions. This may be an indirect indication that the main source of ionization in the 65–85 km region is ionization of excited molecules O_2, rather than of nitric oxide.

All the above concerns the daytime D region at middle latitudes. There is little to add concerning the photochemistry of negative ions at night to Radicella's (1968) work. The suggestion that corpuscular fluxes are the main ionization source in the night D region seems to be reasonable, though other sources such as ionization by scattered $L\alpha$ radiation have been postulated (Tohmatsu and Ogawa, 1966).

The aeronomy of the lower ionosphere at high latitudes is a very complicated question, which requires special consideration. We shall mention only two well-known observations in PCA events: (1) the main negative ion in the D region is not O_2^+, but some ion with high ($\geqslant 3$ eV) electron affinity (Reid, 1964; Reid and Leinbach, 1961); and (2) cosmic noise absorption is strongly dependent on the time of day (Reid, 1964; Bailey, 1959). It seems that the scheme of ionic transformations described above, agrees with the first observations because according to this scheme the main negative ions are NO_2^- or NO_3^-, which have a high electron affinity. The solar control of absorption in PCA events was earlier ascribed to photodetachment, but it is now clear that photodetachment from O_2^+ cannot play any significant role because of the very rapid reactions of O_2^- with O and O_2 ($^1\Delta_g$). Photodetachment from ions with high electron affinity seems to be a weak process. Reid (1969) explained the observed correlation between absorption and proton fluxes during PCA events by the role of process $O_2^- + O$ (reaction 5 in Table IV). It seems however, that the effect of solar control is a result of the detachment reactions both with O and $O_2(^1\Delta_g)$. The latter will give a pronounced variation of $[e]$ from day to night because $O_2(^1\Delta_g)$ concentrations vary strongly after sunset (Gattinger, 1968; Vlasov, 1969).

The photochemistry of the lower ionosphere during aurora is badly known; one may mention recent papers by Donahue et al. (1970) and by Swider and Narcisi (1970), where it was reported, that during aurora there exist additional mechanisms of transformation of O_2^+ to NO^+ in the auroral E region. Donahue (1970, private communication) obtained strong increases of both $[NO^+]/[O_2^+]$ ratio and nitric oxide concentration during bright aurora. It seems to indicate, that reaction (2) is the required mechanism.

4. Metallic Ions

Layers of metallic ions are a phenomenon which is difficult to attribute either to the D or to the E region. In some experiments narrow layers of Fe^+, Mg^+, K^+ and so on were observed at heights above 100 km (Istomin, 1961; Young et al., 1967) that is in the E region. But Narcisi (1966, 1969) also found layers of metallic ions at altitudes about 85 km, in the D region.

The photochemistry of these ions is not yet satisfactorily known. The formation of ions of metals may proceed as a result of photoionization of corresponding neutral atoms M or by ion-molecular reactions with the main atmospheric ions, for example:

$$O_2^+ + M \rightarrow M^+ + O_2 \tag{12}$$

which have high rate coefficients (of the order of 10^{-10}–10^{-11} cm^3 sec^{-1}, Fite, 1969). However, the question of the processes of destruction of M^+ remains open. Ion-molecular reactions with O_3:

$$M^+ + O_3 \rightarrow OM^+ + O_2 \tag{13}$$

can convert metallic ions in molecular ions MO^+ (which have a high recombination coefficient), but the quantity of O_3 in the E region is too small for these reactions to be significant. For M^+ ions, observed at low heights ($h \approx 85$ km) processes of destruction as a result of mutual neutralization with X^- can play a role, especially at night, but nothing is known about its efficiency.

Attempts have been made to explain observed layers of M^+ with the processes of destruction of M^+ of the type (Swider, 1968):

$$M^+ + O \rightarrow OM^+ + h\nu \tag{13}$$

or (Swider, 1969)

$$M^+ + O_2 + Az \rightarrow MO_2^+ + Az \tag{14}$$

and also by slow radiative recombination M^+ together with wind-shear theory (Whitehead, 1966), but a satisfactory solution of the problem has not been achieved yet.

References

Adams, G. W. and Megill, L. R.: 1967, *Planetary Space Sci.* **15**, 1111.
Aikin, A. C.: 1967, in *Ground Based Radio Wave Propagation Studies of the Lower Ionosphere*, Defence Research Board, Ottawa, Canada, p. 18.
Antonova, L. A. and Kazachevskaya, T. V.: 1970, *Space Res.* **10**, North-Holland Publ. Co., Amsterdam, p. 757.
Antonova, L. A., Ivanov-Kholodny, G. S., Kazachevskaya, T. V., Korjagin, A. I., and Medvedev, V. S.: 1970, Report at the XIII-th COSPAR Meeting, Leningrad.
Bailey, D. K.: 1959, *Proc. IRE* **47**, 255.
Barth, C. A.: 1966, *Ann. Geophys.* **22**, 198.
Bates, D. R.: 1956, *J. Atmospheric Terrest. Phys. Suppl.* **6**, 1.
Bowhill, S. A.: 1969, Aeronomy Report N 32, University of Illinois, Urbana, Ill., p. 12.
Danilov, A. D.: 1969, Report at the XII-th COSPAR Meeting, Prague.
Danilov, A. D.: 1970, *Chemistry of the Ionosphere* (ed. by R. W. Fairbridge), Plenum Press, New York.
Danilov, A. D. and Ivanov-Kholodny, G. S.: 1966, *Kosmich. Issled.* **4**, 439.
Danilov, A. D. and Tresvatsky, A. N.: 1970, Report at this Symposium.
Donahue, T. M., Zipf, E. C., and Parkinson, T. D.: 1970, *Planetary Space Sci.* **18**, 171.
Fehsenfeld, F. C. and Ferguson, E. E.: 1969, *J. Geophys. Res.* **74**, 2217.
Fehsenfeld, F. C., Schmeltekopf, A. L., Schiff, H. I., and Ferguson, E. E.: 1967, *Planetary Space Sci.* **15**, 373.
Ferguson, E. E.: 1969a, Report at the IAGA Symposium, Madrid.
Ferguson, E. E.: 1969b, *Can. J. Chem.* **47**, 1815.

Ferguson, E. E. and Fehsenfeld, F. C.: 1969, *J. Geophys. Res.* **74**, 5743.
Ferguson, E. E., Fehsenfeld, F. C., and Schmeltekopf, A. L.: 1967, *Space Res.* **7**, 135.
Fite, W.: 1969, *Can. J. Chem.* **47**, 1797.
Gattinger, R. L.: 1968, *Can. J. Phys.* **46**, 1613.
Holmes, J. C., Johnson, C. Y., and Young, J. M.: 1965, *Space Res.* **5**, 756.
Hunten, D. M. and McElroy, M. B.: 1968, *J. Geophys. Res.* **73**, 2421.
Hunten, D. M. and McElroy, M. B.: 1969, Aeronomy Report N 32, University of Illinois, Urbana, Ill., p. 187.
Inn, C. Y.: 1961, *Planetary Space Sci.* **5**, 77.
Istomin, V. G.: 1961, *Iskysst. Sputniky Zemli* **11**, 98.
Istomin, V. G.: 1963, *Space Res.* **3**, 209.
Ivanov-Kholodny, G. S.: 1964, *Geomagnetizm i Aeronomiya* **4**, 417.
Ivanov-Kholodny, G. S.: 1965, *Space Res.* **5**, 19.
Ivanov-Kholodny, G. S.: 1967, *Dokl. Akad. Nauk SSSR* **177**, 1328.
Ivanov-Kholodny, G. S.: 1969, Report at the IAGA Meeting, Madrid.
Ivanov-Kholodny, G. S. and Kazachevskaya, T. V.: 1966, *Geomagnetizm i Aeronomiya* **6**, 27.
Ivanov-Kholodny, G. S. and Korsunova, L. P.: 1969, *Geomagnetizm i Aeronomiya* **9**, 474.
Ivanov-Kholodny, G. S. and Korsunova, L. P.: 1970, *Geomagnetizm i Aeronomiya* **10**, 532.
Ivanov-Kholodny, G. S. and Nicolsky, G. M.: 1969, *Solntse i Ionosphera*, "Nauka", Moscow.
Kazachevskaya, T. V. and Ivanov-Kholodny, G. S.: 1970, Report at this Symposium.
Korsunova, L. R.: 1966, *Geomagnetizm i Aeronomiya* **6**, 1114.
LeLevier, R. E. and Branscomb, L. M.: 1968, *J. Geophys. Res.* **73**, 27.
Mitra, A. P.: 1959, *J. Geophys. Res.* **64**, 733.
Mitra, A. P.: 1963, *Advances in Upper Atmosphere Research* (ed. by B. Landmark), Pergamon Press, p. 57.
Mitra, A. P.: 1969a, *J. Atmospheric Terrest. Phys.* **30**, 1065.
Mitra, A. P.: 1969b, Aeronomy Report N 32, University of Illinois, Urbana, Ill., p. 174.
Moler, W. F.: 1960, *J. Geophys. Res.* **65**, 1459.
Narcisi, R. S.: 1966, *Ann. Geophys.* **22**, 224.
Narcisi, R. S.: 1967, *Space Res.* **7**, 186.
Narcisi, R. S.: 1969, *Space Res.* **8**, 360.
Nicolet, M. and Aikin, A. C.: 1960, *J. Geophys. Res.* **65**, 1469.
Pearce, J. B.: 1969, *J. Geophys. Res.* **74**, 853.
Phelps, A. V.: 1969, *Can. J. Chem.* **47**, 1783.
Pontano, B. A. and Hale, L. C.: 1970, *Space Res.* **10**, in press.
Radicella, S. M.: 1968, *J. Atmospheric Terrest. Phys.* **30**, 1745.
Ratcliffe, J. A. and Weekes, K.: 1960, *Physics of the Upper Atmosphere* (ed. by J. A. Ratcliffe), Academic Press, New York and London.
Reid, G. C.: 1964, *Rev. Geophys.* **2**, 311.
Reid, G. C.: 1969, *Planetary Space Sci.* **17**, 731.
Reid, G. C. and Leinbach, H.: 1961, *J. Atmospheric Terrest. Phys.* **23**, 216.
Schoen, R. E.: 1969, *Can. J. Chem.* **47**, 1879.
Serafimov, K. B.: 1970, *Ionosph. Issled.* **19**, 99.
Swider, W.: 1965, *J. Geophys. Res.* **70**, 4859.
Swider, W.: 1968, *Nature* **217**, 438.
Swider, W.: 1969, Report at the IAGA Symposium, Madrid.
Swider, W. and Keneshea, T. J.: 1968, *Space Res.* **8**, 370.
Swider, W. and Narcisi, R. S.: 1970, *Planetary Space Sci.* **18**, 379.
Tohmatsu, T. and Ogawa, T.: 1966, *Rept. Ionospheric Space Res. Japan* **20**, 395.
Tulinov, G. F.: 1967, *Space Res.* **7**, 386.
Vlasov, M. N.: 1969, *Geomagnetizm i Aeronomiya* **9**, 940.
Vlasov, M. N.: 1970, Report at this Symposium.
Whitehead, J. D.: 1966, *Radio Sci.* **1**, 198.
Whitten, R. C. and Poppoff, I. G.: 1964, *J. Atmospheric Sci.* **21**, 117.
Whitten, R. C. and Poppoff, I. G.: 1965, *Physics of the Lower Ionosphere*, Prentice Hall.
Yonezawa, T.: 1968, *J. Atmospheric Terrest. Phys.* **30**, 473.
Young, J. M., Johnson, C. Y. and Holmes, J. C.: 1967, *J. Geophys. Res.* **72**, 1473.

THE CHEMISTRY OF NEUTRAL SPECIES
IN THE D AND E REGIONS

S. A. BOWHILL

Aeronomy Laboratory, Dept. of Electrical Engineering, University of Illinois, Urbana, Ill. 61801, U.S.A.

Abstract. The photodissociation of molecular species such as oxygen, nitrogen, and water vapor in the upper atmosphere leads to the formation of a reservoir of chemical energy consisting of highly reactive atomic gases. The reactions of these with each other, and with the rest of the atmosphere, gives a complex neutral chemistry involving many species and reactions. Local equilibrium can rarely be assumed for these reactions; transport by the horizontal and vertical mean flow, by eddy diffusion, and by molecular diffusion has quite different effects on the various constituents, depending on the time constants. The entire subject of the neutral chemistry in the D and E regions (between 50 and 160 km altitude) is reviewed, and related to existing experimental results. Needs for further research on the subject are indicated.

1. Introduction

The chemistry of the neutral species of the D and E regions of the ionosphere is controlled by three separate types of process: chemical modification of the atmosphere by solar radiation, transport effects, and chemical interactions of the various species. Each of these processes is of considerable complexity, and only in the past few years has it been possible to construct realistic models for them. In part, the difficulty of constructing realistic theories results from a lack of reliable observational data for the densities of many of these species, and of their variability with time, place and season.

In Section 2 of this paper, the neutral species which enter into the chemistry of the D and E regions are reviewed, together with the role of solar radiation in chemically modifying them. Section 3 describes the effect of the field of the atmosphere which (together with molecular diffusion) is responsible for redistributing the neutral species in the vertical and horizontal directions. Section 4 reviews the production, redistribution and disappearance of each significant constituent in turn, indicating the significance of the experimental evidence for its morphology. Finally, Section 5 gives a list of outstanding problems in the neutral chemistry of the D and E regions, where additional theoretical or experimental research can best be deployed.

2. Neutral Species and their Dissociation

At the low pressures prevailing in the D and E regions of the ionosphere, many neutral gases are found which are so reactive that they do not appear in the lower atmosphere. Most of the reactions that occur involve the exothermic transfer of a loosely bound oxygen, hydrogen or nitrogen atom from one particle to another as a simple bimolecular reaction, for example:

$$O_3 + H \rightarrow O_2 + OH + 76.8 \text{ kcal}. \tag{1}$$

Dyer (ed.), Solar-Terrestrial Physics/1970: Part IV, 41–52. *All Rights Reserved.*

This reaction is particularly important as the enthalpy decrease can appear as excitation of the OH radical up to the ninth vibrational level, giving rise to the OH airglow.

The following are the principal species of molecules with available oxygen atoms that appear in the D and E regions (in ascending order of oxygen atom binding energy): O_3, H_2O_2, NO_3, HO_2, NO_2, HNO_3, HNO_2, OH, O_2, CO_2 and NO. Species with available hydrogen atoms (also in increasing order of binding energy) are: HO_2, HNO, HNO_2, H_2O_2, HNO_3, CH_4, OH, H_2 and H_2O. For nitrogen atoms, the species are: NO_2, N_2O, NO and N_2.

All of these molecules are subject to dissociation by solar radiation; in most cases, the effect being to remove the available oxygen, hydrogen or nitrogen atom. In some cases, visible light can dissociate the molecule, as in:

$$NO_2 + hv\,(2900\text{--}4400\ \text{Å}) \rightarrow NO + O \tag{2}$$

while in other cases, the absorption spectrum of the molecule has several bands, giving different dissociation products. For example O_2 can be dissociated in the Schumann-Runge continuum (<1750 Å) to give:

$$O_2 + hv \rightarrow O\,(^1D) + O\,(^3P) \tag{3}$$

or in the Schumann-Runge bands and Herzberg continuum to give:

$$O_2 + hv \rightarrow O\,(^3P) + O\,(^3P) \tag{4}$$

For dissociation to occur to a significant extent by a given wavelength of solar radiation, two conditions need to be fulfilled. Firstly, the absorption cross section of the molecular species must be great enough that the dissociation proceeds at least as rapidly as the competing influences of chemical reaction and transport. Secondly, the solar radiation must penetrate to the altitude in question, bearing in mind that a number of molecular species may contribute to the absorption.

It is convenient to use two parameters for each constituent to represent these two criteria. The first is the lifetime τ_d against dissociation, defined as the reciprocal of the dissociation rate coefficient. This represents the time for 63% of the species to be dissociated by solar radiation, in the absence of any competing process. If this lifetime is greater than about one day in the D and lower E regions, eddy transport effects will be important. If the lifetime is much shorter than one day, it is likely that the rate dissociation is nearly balanced by the chemical loss processes, at least in the middle of the day; and that there is substantial diurnal variation in the density of the dissociated constituent. Molecular diffusion becomes important above about 105 km, and further lowers the transport time (see Section 3).

The second parameter is the altitude z_d of unit optical depth in the atmosphere for vertically incident radiation which dissociates a given constituent; namely, the altitude at which only 37% of the incident solar intensity still remains, for an overhead sun. A few kilometers above that altitude, negligible attenuation of the dissociating radiation has occurred; below that altitude, the solar radiation intensity decreases extremely rapidly. Table I gives values for τ_d and z_d.

TABLE I

Unit optical depth altitudes, z_d and dissociation time
constants τ_d for the important molecular species

Constituent dissociated	Wavelengths (Å)	z_d (km)	τ_d
NO_2	2900–4400	0	180 sec
O_3	<3100	45	200 sec
	4500–7500	0	1 hr
HO_2	<3100	~45	~1 hr
H_2O_2	<3100	45	2 hr
	>3100	0	10 hr
CH_4	1216	75	12 hr
H_2O	1216	75	2 days
	1724–7325	~65	~20 days
O_2	<1750	110	2 days
	1950–2424	45	30 yr
CO_2	<1690	100	6 days
N_2O	<3000	45	20 days
	3000–3370	0	60 yr
NO	~1900	45	110 days

Dissociation can be completely neglected for N_2, OH and H_2, all of which have negligible dissociation cross sections for the significant radiation.

Inspection of Table I shows that transport can be neglected in determining the distributions of NO_2, O_3, HO_2 and H_2O_2, which are therefore controlled by photochemistry alone. However, transport plays a substantial or dominant role for the constituents CH_4, N_2O, O_2, CO_2, H_2O and NO.

As will be shown in Section 4, chemical reactions in the absence of solar radiation rapidly produce oxidizing species (NO_2, O_3, HO_2) from those atoms (O, H) which have a reducing action. Photodissociation acts to reverse this tendency, and produce a reducing atmosphere. The spectacular effect of solar radiation in changing an oxidizing atmosphere into a reducing atmosphere is shown in Figure 1 (Hunt, 1966), representing theoretical calculations for an altitude of 76 km in the D region. During the night a gradual decrease takes place in the reduceants $O(^3P)$ and H, with a complementary increase in O_3 and HO_2. As soon as the sun rises, however, the situation is completely reversed, each of the oxidants producing a reduceant in almost the same concentration, by the process of photodissociation. Obviously, one must expect dramatic changes in the ion chemistry to accompany this situation.

Of course, each of the neutral constituents has its own structure of energy levels; relatively simple in the case of the atomic species, but including electronically and vibrationally excited states for molecules. These excited species have particular aeronomic significance for two reasons. Firstly, radiation by these species is responsible for the phenomenon of airglow; and the complexity of the airglow spectrum reflects the degree of complexity of the excited species composition of the upper atmosphere at the time of observation. A second reason why these excited species have particular significance is that they act as a reservoir of additional energy which the atom or molecule

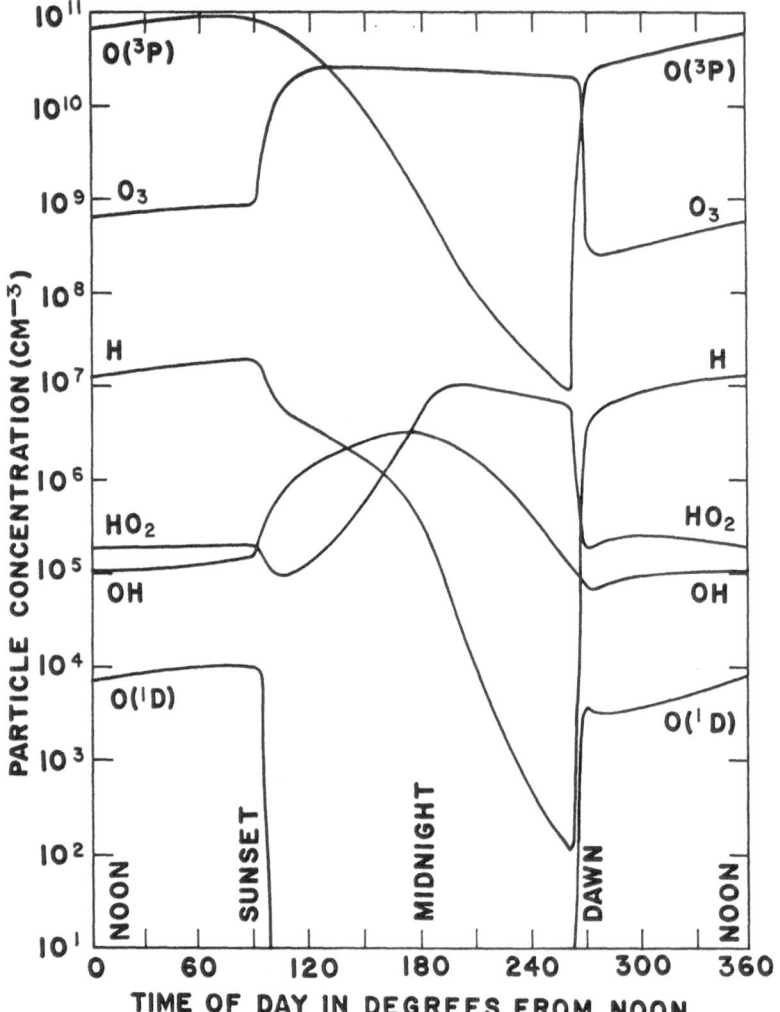

Fig. 1.   Calculated composition at 76 km (Hunt, 1966).

can use to enable a chemical reaction to occur which would otherwise not be energetically possible, or which might have a prohibitively high activition energy.

The most significant excited species in the D and E regions are the (^1S) and (^1D) states of atomic oxygen; the (^2D) state of atomic nitrogen; the ($^1\Delta$) and ($^1\Sigma$) states of molecular oxygen; and the vibrationally excited states of molecular nitrogen and hydroxyl. Their behavior has been reviewed in detail by Dalgarno (1970).

3. Effects of the Motion Field of the Atmosphere

Two aspects of atmospheric motion are important in considering the neutral chemistry of the D and E regions. If the atmosphere is assumed to be static, photochemical processes would produce substantial changes in concentration of various constitutents in distances as small as 1 km in height. Obviously, if the constituent is sufficiently long-lived that it can be moved vertically while still retaining its identity, the static atmosphere solution cannot be applied.

While our knowledge of the gross motions of the atmosphere above 50 km is in a far from satisfactory state, it is clear that at least four types of motion can affect the neutral composition. The most significant of these in the E region is molecular diffu-sion, in which each constituent tends to adopt an altitude distrubution as though it were in hydrostatic equilibrium with its own scale height. The rapidity with which this process takes place increases exponentially with altitude, and the diffusion coefficient can be easily calculated for any neutral constituent. Below about 105 km, however, the so-called 'turbopause', the process of eddy diffusion dominates that of molecular diffu-sion. The principal difference between these two processes is that the eddy diffusion tries to produce a mixing distribution for a minor constituent; namely, a distribution in which the minor constituent concentration is a constant fraction of the total particle concentration. While an eddy diffusion equation can be written, similar to the mole-cular diffusion equation, it must be said that the basic concept of eddy diffusion is not on very firm theoretical ground. In part, this results from our ignorance of the precise mechanism which produces it, whether from a kind of turbulence, or the effects of systems of gravity wave motions (Hodges, 1969). Nevertheless, it has become conven-tional to adopt the eddy diffusion model as a convenient working representation, even though the height variation of the eddy diffusion coefficient is virtually unknown.

Two other types of motion have been suggested as important for the convection of minor constituents. Christie (1970) has discussed the importance of planetary waves, and has suggested that they can convect minor constituents along meridional isen-tropic contours (Figure 2). Large-scale meridional circulation patterns, with their associated subsidence motions (Geisler and Dickinson, 1968) also can produce strong horizontal convergence.

Ultimately, each of these four types of motion will need to be represented rigorously in minor constituent transport equations. However, at present, it seems that a combi-nation of eddy and molecular diffusion is as good an approximation as is warranted, given our present state of knowledge about the atmospheric chemistry.

Generally speaking, transport effects for relatively long-lived constituents under eddy diffusion involve three possible regions; a source region, a region where the constituent is being convected by eddy diffusion, and a sink region. The source or sink may be photochemical, or may be the large reservoir represented by the lower atmosphere.

The one-dimensional eddy diffusion equation obeyed by a minor constituent of concentration N under gravity is

$$G = - D \left(\frac{dN}{dz} + \frac{N}{H} \right) \tag{5}$$

where G is the flux of the constituent in the direction of positive altitude z, and H is the scale height of the atmosphere (H and D are supposed constant in this simple example). In the diffusion region between the source and sink, the continuity equation is just $dG/dz=0$. The solution to Equation (5) is therefore:

$$N = Ae^{-z/H} - \frac{G_0 H}{D} \tag{6}$$

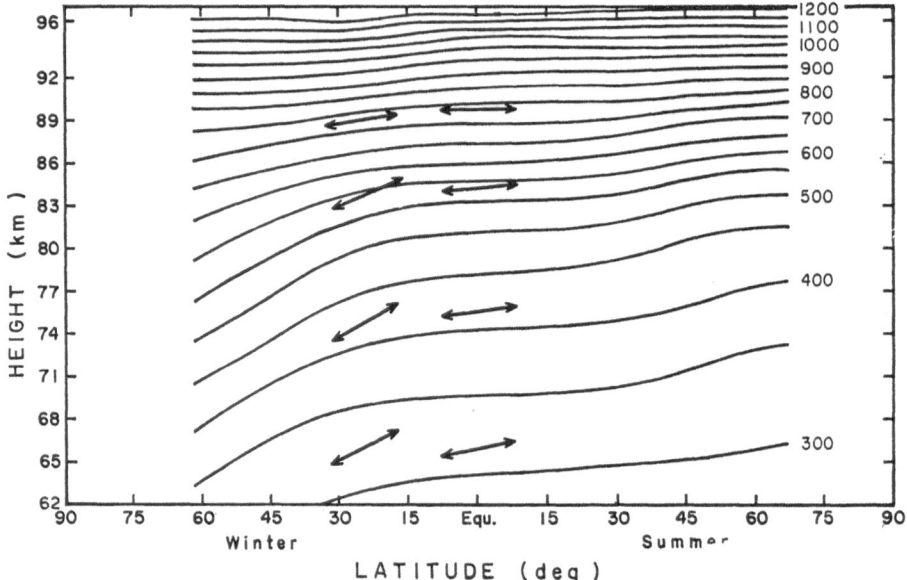

Fig. 2.   Possible meridional flow for minor constituents (Christie, 1970).

where G_0 is the constant flux in the diffusion region, and A is a constant. If the sink, located at a level z_1, is ideal it imposes the boundary condition $N=0$ at $z=z_1$; then

$$N = \frac{G_0 H}{D} (e^{(z_1-z)/H} - 1) \quad (z \leqslant z_1) \tag{7a}$$

or

$$N = \frac{(-G_0) H}{D} (1 - e^{(-z-z_1)/H}) \quad (z \geqslant z_1) \tag{7b}$$

where Equation (7a) refers to an upward flux, with the sink located over the source, while (7b) applies to a downward flux with the source located over the sink. These expressions show that for the sink-over-source condition (7a), the mixing ratio $Ne^{z/H}$ is nearly constant, at the value near the source, to within one or two scale heights of the sink; while in the source-over-sink condition (7b), the density N is nearly constant at its source value to within one or two scale heights of the sink.

Table II illustrates how various atmospheric species are controlled by the locations of their sources and sinks. The first two (O and NO) represent a source-over-sink configuration, and have a density in the eddy diffusion region which (by Equation

TABLE II

Species	Source height	Source type	Sink height	Sink type
O	105–115 km	Dissociation	75–90 km	Chemical
NO	> 100	Chemical	< 20	Chemical
H_2O	0	Troposphere	70–85	Dissociation
N_2O	0	Troposphere	40–60	Dissociation
CO	0	Troposphere	~ 30	Chemical
CH_4	0	Troposphere	~ 30	Chemical

7b)) is approximately inversely proportional to the eddy diffusion coefficient. Figure 3, due to Colegrove *et al.* (1966) shows the results of numerical solution to the atomic oxygen transport problem, showing that this inverse relationship is obeyed rather exactly. This result is specially significant in that a variation of the eddy diffusion coefficient between the source and the sink will result in a change in the atomic oxygen concentration, not only at 120 km, but at all higher altitudes as well, including the region above 200 km where the atmosphere is dominantly atomic oxygen.

The remaining four constituents in Table II (H_2O, N_2O, CO and CH_4) have a sink-over-source arrangement, and so, by Equation (7a) have nearly constant mixing ratios up to the sink region.

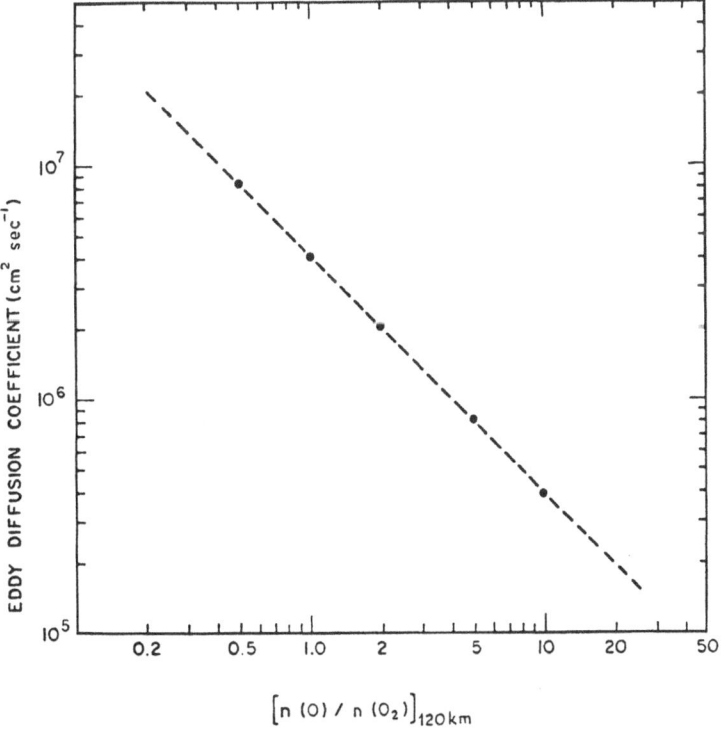

Fig. 3. Eddy diffusion effect on composition (Colegrove *et al.*, 1966).

4. Chemistry of Neutral Species in the D and E Region

The many complicated chemical reactions which can take place between the simple compounds of nitrogen, oxygen and hydrogen have been reviewed by Schofield (1967), Kaufman (1969), Schiff (1969) and Nicolet (1964, 1965, 1970). Many of the significant reaction rates are now known with some accuracy. However, combining them in a comprehensive way, including all relevant transport effects, has proved to be exceedingly difficult, because of the wide range of time constants involved. The work of Hunt (1966), the first to include time-dependent solutions for ozone and hydrogen compounds, neglected transport altogether; Hesstvedt (1968) used implicit forms for the concentrations of the short-lived constituents, combined with transport for those with a longer life; and Shimazaki and Laird (1970) used a combination of eddy and molecular diffusion for all constituents.

Experimental evidence regarding atomic oxygen concentrations comes from rocket-borne mass spectrometers; although the measurement is difficult to make because of recombination of atomic oxygen on the walls of the spectrometer, there seems general agreement that the concentrations of atomic and molecular oxygen are equal at an altitude of about 120 km. Below 100 km, unfortunately, mass spectrometers cannot be used, and no satisfactory alternative technique is available.

The dynamics of the atomic oxygen-ozone system are dominated by the competing processes of ozone photolysis and three-body recombination of atomic oxygen:

$$O_3 + hv \rightarrow O + O_2 \tag{8}$$
$$O + O_2 + M \rightarrow O_3 + M + 25.4 \text{ kcal}. \tag{9}$$

These two reactions leave the total concentration of 'odd' oxygen unchanged, so much attention has been given to mechanisms by which odd oxygen can be destroyed. In a dry atmosphere (hydrogen compounds absent), the significant reactions are:

$$O + O + M \rightarrow O_2 + M + 25.6 \text{ kcal} \tag{10}$$
and
$$O + O_3 \rightarrow 2O_2 + 93.7 \text{ kcal}. \tag{11}$$

However, the first is relatively slow if the atomic oxygen concentration is small, and the second has a significant (4.0 kcal) activation energy. Observations of the ozone concentration above 60 km (Weeks and Smith, 1968; Anderson et al., 1969; Hilsenrath et al., 1969) show a much lower concentration of ozone than would be calculated for a dry atmosphere. Hunt (1966) carried out non-equilibrium calculations involving atomic hydrogen in a catalytic role, the principal reactions being:

$$O_3 + H \rightarrow O_2 + OH + 76.8 \text{ kcal} \tag{1}$$
$$OH + O \rightarrow H + O_2 + 16.9 \text{ kcal} \tag{12}$$
$$H + O + M \rightarrow HO_2 + M + 107 \text{ kcal} \tag{13}$$
$$HO_2 + O \rightarrow OH + O_2 + 55 \text{ kcal}. \tag{14}$$

However, his principal source for hydrogen compounds was $O(^1D)$:

$$O\,(^1D) + H_2O \rightarrow 2OH + 26.1 \text{ kcal} \tag{15}$$

and the quenching coefficient for $O(^1D)$ has been found to be much greater, and the concentration much less, than the value he used. The principal D-region source of active hydrogen compounds is therefore photolysis of water vapor:

$$H_2O + hv \rightarrow H + OH \tag{16}$$

which is re-formed relatively slowly in the mesosphere. The active hydrogen compounds (H, OH, HO_2) are continually oxidized and reduced by the reactions (1), (12), (13) and (14) above, together with:

$$H + O_2 + M \rightarrow HO_2 + M + 47 \text{ kcal} \tag{17}$$
$$OH + O_3 \rightarrow HO_2 + O_2 + 39 \text{ kcal}, \tag{18}$$

the latter reaction being significant chiefly at night, when the disappearance of H causes reaction (17) to become insignificant. Most of the active hydrogen is lost to H_2:

$$H + HO_2 \rightarrow H_2 + O_2 + 57 \text{ kcal}. \tag{19}$$

The H_2 so formed has a long lifetime, and diffuses to above 100 km, where it is presumably dissociated and becomes the main source of the geocoronal hydrogen (Bates and Nicolet, 1965; Brinkmann, 1969).

As a consequence of its loss by dissociation, water vapor is continuously transported upwards (see Table II), its mixing ratio staying approximately constant till about 70 km altitude, where it begins to drop rapidly. Recent calculations by Shimazaki and Laird (1970) have shown that significant errors can arise in neglecting the transport effects in the D region for both oxygen and hydrogen compounds.

Great interest has developed recently in the chemistry of the oxides of nitrogen. The rapid processes, analogous to (14) and (18):

$$NO_2 + O \rightarrow NO + O_2 + 45.9 \text{ kcal} \tag{20}$$
$$NO + O_3 \rightarrow NO_2 + O_2 + 47.8 \text{ kcal} \tag{21}$$

together with the rapid photolysis of NO_2 (see Table I), means that 'odd' nitrogen is primarily in the form of NO throughout the D region in daytime but its oxidation to NO_2 causes the latter to predominate below about 70 km at night.

The mechanism for production of nitric oxide from nitrogen atoms was long thought to be (Nicolet, 1965)

$$N + O_2 \rightarrow NO + O + 91.4 \text{ kcal}, \tag{22}$$

with a 7.1 kcal activation energy, which was then balanced by:

$$N + NO \rightarrow N_2 + O + 134.6 \text{ kcal}. \tag{23}$$

This pair of reactions has the property that the resulting concentration of NO is independent of the concentration of N, if chemical equilibrium prevails; it gives the concentrations shown in the first column of Table III. Rocket measurements of the NO concentration by Barth (1966) and by Pearce (1969) from the fluorescence of the γ-bands (also shown in the table) gave much larger values; Pearce's values, indeed, being so large as to pose great difficulties in reconciling them with the observed electron density in the D region. Very recent observations by Meira (private communication, 1970) have pointed to lower values, and it is believed that the reason for the higher values of Pearce is now understood.

TABLE III

Comparison of observed and calculated D-region nitric oxide concentrations

Altitude (km)	Nicolet (1965)	Barth (1966)	Pearce (1969)	Meira (1970)	Strobel et al. (1970)
90	2×10^5 cm^{-3}	5×10^7	10^8	2×10^7	5×10^7
80	1.4×10^6	5×10^7	2×10^8	1.5×10^7	10^8
70	5×10^6	–	10^9	6×10^7	

In any case, another source of NO is evidently required. Norton (1967) suggested that excited N could be involved:

$$N(^2D) + O_2 \rightarrow NO + O + 144 \text{ kcal} \tag{24}$$

which is rapid and unlike (22), has no activation energy. The consequences of this reaction have been explored by Norton and Barth (1970) and by Strobel et al. (1970), who find that it is generally adequate to match both D- and E-region measurements of NO concentration (Table III).

The long lifetime of odd nitrogen (that is, N, NO and NO$_2$ combined) had been noted by Bates and Hays (1967) in connection with their studies of N$_2$O. Produced by reaction (24) principally at altitudes above 100 km, the odd nitrogen slowly descends under the influence of eddy diffusion until it is washed away in the troposphere. Since this is a classical example of a 'source-over-sink' situation, one would expect a nearly constant concentration of NO with altitude in the D region; the concentration in fact varying only if the eddy diffusion coefficient varies. If the current theory is substantiated, therefore, it appears that measurement of nitric oxide concentration may give the best index of eddy diffusion rates in the D region.

Nitrous oxide (N$_2$O) has been discussed in detail by Bates and Hays (1967). It is produced at the surface of the earth, and dissociated in the stratosphere. Its relevance to the D region is that it provides a background mixing ratio of odd nitrogen atoms that could dominate the NO concentration in the lowest part of the D region.

Finally, there are a number of highly significant excited species in the D and E region, which are listed in Table IV. The 'crossover altitude' is where the lifetime against quenching is equal to the radiative lifetime for the constituent. Their chemistry has been recently discussed exhaustively by Zipf (1969) and by Dalgarno (1970) and little can be added here. Their lifetimes are all sufficiently short that transport effects

TABLE IV

Principal electronically excited species in D and E regions

Species	Energy excess (eV)	Radiative lifetime (sec)	Quenching agent	Average coefficient (cm^3 sec^{-1})	Crossover altitude (km)
$O(^1S)$	4.1	0.74	O_2	2×10^{-13}	94
$O(^1D)$	1.9	110	N_2, O_2	7×10^{-11}	> 200
$N(^2D)$	2.3	9×10^4	O_2	10^{-12}	> 300
$O_2(^1\Delta)$	1.1	3×10^3	N_2	3×10^{-19}	70
$O_2(^1\Sigma)$	2.0	12	N_2	1.5×10^{-15}	90

can generally be neglected in the D and E regions. They are generally produced by photolysis; by resonance-phosphorescence; by dissociative recombination of positive ions and electrons; by 2-body or 3-body chemical reactions where the enthalpy is adequate (considering that 1 eV $\equiv$ 23.1 kcal mole^{-1}); and by electron impact. All are detectable in the day airglow, and some at night, but mainly near or above their crossover altitudes.

Vibrationally excited N_2 is an exception in that its transport is significant (Walker *et al.*, 1969). It may be important in heating E-region electrons through superelastic collisions, and in its effect on the ion-atom interchange reaction between N_2 and atomic oxygen ions.

5. Problems in D and E Region Neutral Chemistry

There are four general areas for research in this rapidly developing subject.

Firstly, there is a need for the development of new experimental techniques for the measurement of oxidants and reducants in the D region. At the moment it is necessary to rely on theoretical inferences from laboratory rate coefficients to a deplorable extent.

Secondly, better laboratory measurements are needed for gaseous reactions of aeronomic interest, including careful study of the state of excitation of the reactants and reaction products. Primarily, the requirements here are to draw the needs to the attention of the large number of atomic collision scientists who would be competent to research on this problem.

Thirdly, careful study is needed of the meteorological transport process which can affect the concentrations of long-lived neutral constituents. New techniques are required for the measurement of eddy diffusion coefficients and mass transport rates. Failing such techniques, it will be necessary to use neutral species as tracers to determine the nature of the motion field; for example comparing water vapor, primarily transported upward, with the nitrogen oxides, primarily transported downward. Radioactive tracers and ozone distributions can also be exploited for this purpose.

Fourthly, in planning experimental programs to study the aeronomy of the D and E regions, careful attention should be given to the relationship between neutral and ion-chemical effects. The past few years have taught us that the D and E regions are not the static medium, governed by one or two continuity equations, which we had previously thought them; but an extremely complex region where the energy of solar photolysis interacts with the tropospherically generated motion field of the upper atmosphere.

Acknowledgement

Preparation of this paper was supported by the National Aeronautics and Space Administration under Grant NGR14-005-013.

References

Anderson, G. P., Barth, C. A., Cayla, F., and London, J.: 1969, *Ann. Geophys.* **25**, 341.
Barth, C. A.: 1966, *Ann. Geophys.* **22**, 198.
Bates, D. R. and Hays, P. B.: 1967, *Planetary Space Sci.* **15**, 189.
Bates, D. R. and Nicolet, M.: 1965, *Planetary Space Sci.* **13**, 905.
Brinkmann, R. T.: 1969, *J. Geophys. Res.* **75**, 5335.
Christie, A. D.: 1970, *J. Atmospheric Terrest. Phys.* **32**, 35.
Colegrove, F. D., Johnson, F. S., and Hanson, W. B.: 1966, *J. Geophys. Res.* **71**, 2227.
Dalgarno, A.: 1970, *Ann. Geophys.* **26**, 601.
Geisler, J. E. and Dickinson, R. E.: 1968, *J. Atmospheric Terrest. Phys.* **30**, 1505.
Hesstvedt, E.: 1968, *Geophys. Publ.* **24**, No. 4.
Hilsenrath, E., Seiden, L., and Goodman, P.: 1969, *J. Geophys. Res.* **74**, 6873.
Hodges, R. R.: 1969, *J. Geophys. Res.* **74**, 4087.
Hunt, B. G.: 1966, *J. Geophys. Res.* **71**, 1385.
Kaufman, F.: 1969, *Can. J. Chem.* **47**, 1917.
Nicolet, M.: 1964, *Disc. Faraday Soc.* **37**, 7.
Nicolet, M.: 1965, *J. Geophys. Res.* **70**, 679.
Nicolet, M.: 1970, *Ann. Geophys.* **26**, 531.
Norton, R. B.: 1967, Ph.D. Thesis, University of Colorado, Boulder.
Norton, R. B. and Barth, C. A.: 1970, *J. Geophys. Res.* **75**, 3903.
Pearce, J. B.: 1969, *J. Geophys. Res.* **74**, 853.
Schiff, H. I.: 1969, *Can. J. Chem.* **47**, 1903.
Schofield, K.: 1967, *Planetary Space Sci.* **15**, 643.
Shimazaki, T. and Laird, A. R.: 1970, *J. Geophys. Res.* **75**, 3221.
Strobel, D. F., Hunten, D. M., and McElroy, M. B.: 1970, *J. Geophys. Res.* **75**, 4307.
Walker, J. C. G., Stolarski, R. S., and Nagy, A. F.: 1969, *Ann. Geophys.* **25**, 831.
Weeks, L. H. and Smith, L. G.: 1968, *Planetary Space Sci.* **16**, 1189.
Zipf, E. C.: 1969, *Can. J. Chem.* **47**, 1863.

TRANSPORT PROCESSES IN THE THERMOSPHERE

FRANCIS S. JOHNSON

The University of Texas at Dallas

and

Space Sciences Center, Southern Methodist University, Dallas, Tex., U.S.A.

Abstract. The atmosphere is highly stratified, even to high levels, and hence important transport questions arise in which vertical transport alone constitutes a useful approximation. Thus, the thermosphere was first described in terms of molecular conduction transferring heat from higher levels to lower, which is required because of the lack of adequate energy emission mechanisms at the higher levels. Later, the importance of eddy conduction was recognized as a means of extending the treatment well down into the mesosphere, where finally radiative processes are powerful enough to dispose of the excess heat deposited at the higher levels. Transport problems also arise with regard to atmospheric constituents, and several studies have been made of this problem. These considerations have provided a conceptually complete description of the upper atmosphere in the approximation that horizontal transport can be neglected. Two different approaches have been undertaken to consider the horizontal transport by large-scale circulation. First and most direct, the horizontal wind field in the middle and upper thermosphere was calculated based on the pressure distributions derived from satellite orbital decay observations. Then continuity conditions were used to calculate the vertical motions. The second approach was to evaluate the energy deficits or surpluses as a function of altitude and latitude and to assume the presence of vertical motions sufficient to balance these deficits or surpluses by compressional heating or cooling. Then continuity conditions were used to derive horizontal winds. The results of the two approaches are compatible, and are largely complementary. Downward velocities are near 1 m s^{-1} at 300 km over the diurnal minimum and the winter polar region, and 1 cm s^{-1} at 100 km over the winter polar region.

1. Introduction

The atmosphere is highly stratified at all levels. A considerable degree of understanding of the vertical structure of the atmosphere can be attained by considering vertical transport processes alone. This is not to say that horizontal variations and horizontal transport are not important – they are – but if a comprehensive understanding of the structure of the atmosphere is to be developed in steps it is necessary first to understand the effects of the vertical transport processes.

One of the earliest studies contributing to such understanding was that of Emden (1913), who explained in terms of radiation equilibrium and eddy mixing the division of the atmosphere into troposphere and stratosphere. Later Gowan (1936) showed how the ozone layer perturbed the radiation balance at higher levels to produce a warm region in the atmosphere near 50 km. Spitzer (1949) showed how a hitherto neglected transport process – molecular conduction – controlled the temperature distribution of the outer portion of the atmosphere and caused it to become nearly isothermal at the highest altitudes. It had been recognized earlier from ionospheric data and hydrostatic considerations that the atmosphere above 100 km was quite warm.

Dyer (ed.), Solar-Terrestrial Physics/1970: Part IV, 53–67. All Rights Reserved.
Copyright © 1972 by D. Reidel Publishing Company.

Johnson and Wilkins (1965) identified the important role of eddy transport as the dominant heat transfer process in the altitude region 60–110 km, above which molecular conduction and below which radiation are the dominant energy transfer processes. This filled an important conceptual gap in the understanding of the thermal structure of the atmosphere.

A similar pattern of development occurred with regard to transport of atmospheric constituents. It was recognized at an early time that the effect of molecular diffusion was such as to cause each atmospheric constituent to tend to distribute itself in the vertical as if no others were present. On this account, hydrogen was once thought to be the predominant constituent of the upper atmosphere (Jeans, 1916) because of its low mass. However, Epstein (1932) recognized that molecular diffusion proceeded so slowly that diffusion equilibrium was not likely to become established because mixing would occur too frequently to permit it to be attained. Nicolet (1954) interpreted measurements of atomic and molecular oxygen as indicating that both mixing and molecular diffusion were important near 110 km. Jacchia's (1961) measurements of the variation with altitude of satellite drag were interpreted by Nicolet (1961) as indicating that helium was the predominant constituent of the atmosphere above some high altitude, roughly the lower exosphere. This provided clear evidence that the helium distribution was in diffusive equilibrium and that molecular diffusion was therefore the predominant process controlling the helium distribution above about 110 km, thus implying that mixing was predominant at lower altitudes. Colegrove et al. (1965, 1966) solved the continuity problem for molecular and atomic oxygen and found that this imposed the requirement of a rate of eddy mixing near $4.5 \times 10^6 \text{ cm}^2 \text{ s}^{-1}$ to provide the necessary transport.

After an understanding of the vertical structure of the upper atmosphere has been developed, the next step should be to examine horizontal transport and to develop an understanding of diurnal and latitudinal variations. This has not yet reached the same level of understanding as the vertical structure problem. Kohl and King (1967) have calculated the diurnal winds to be expected at the equinox due to the diurnal pressure gradients in the upper atmosphere near 300 km. Rishbeth et al. (1969) have gone a step further and calculated the pattern of vertical winds to be expected from the horizontal divergence in the calculated horizontal winds. Johnson and Gottlieb (1970) have studied the heat budget of the lower thermosphere during solstice conditions and have calculated the meridional wind distribution required to balance the lack of symmetry in the solar heat input. Fortunately all these results appear reasonably in consonance and supplementary in nature. However, one major observational feature of the upper atmospheric pattern of density and temperature variation has not been adequately explained. Satellite drag data have shown that the maximum of atmospheric density occurs near 1400 hours local time (Jacchia, 1964), whereas calculations of temperature variations to be expected neglecting horizontal transport indicate that the maximum should occur near 1700 hours. It is reasonable to expect that transport by wind systems and associated broad scale vertical currents will explain this discrepancy, but this has not yet been accomplished.

2. Thermal Transport in the Vertical

Figure 1 shows typical rates of heat input into the mesosphere and the lower thermosphere over the equator at solstice (Johnson and Gottlieb, 1970). The curve labeled O_2 represents the heat input above given altitudes due to the absorption of solar ultraviolet radiation by molecular oxygen; the values are averages over a day, obtained by adding hourly values throughout the day and then dividing by 24 hours.

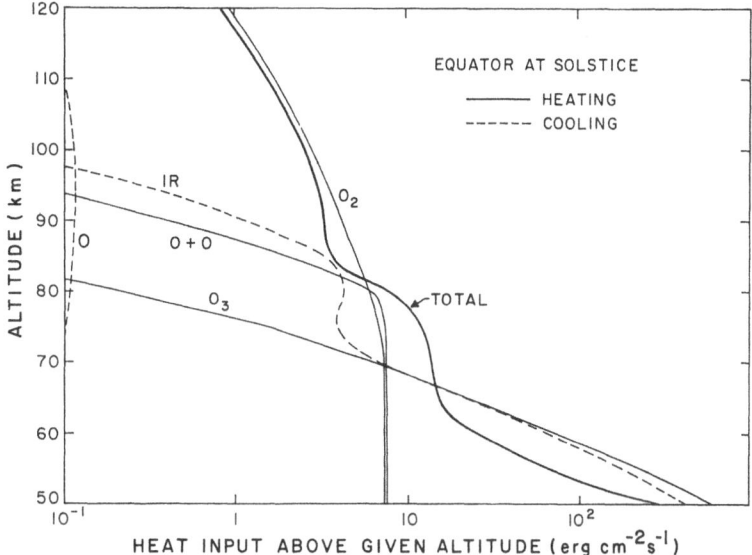

Fig. 1. Heat input into the upper atmosphere above given altitudes evaluated over the equator at the solstice, according to Johnson and Gottlieb (1970). O_2 indicates heat released due to absorption by molecular oxygen of solar radiation in the wavelength range below 1875 Å; O_3 that due to absorption of solar radiation by ozone: $O + O$ the heat release due to recombination of atomic into molecular oxygen; O the heat loss by infrared emission by atomic oxygen according to Craig and Gille (1969); IR the infrared losses due to other constituents according to Kuhn and London (1969).

The curve labeled $O+O$ represents the heat released above given altitudes by the recombination of atomic oxygen into molecular form; this occurs predominantly at a much lower altitude than where the photodissociation is strongest; a transport of molecular oxygen upward and atomic oxygen downward is involved in order to maintain continuity in the steady state distributions. The O_3 curve in Figure 1 represents the heat released due to absorption of solar ultraviolet radiation by ozone. The O curve shows the heat loss above given altitudes due to infrared radiation by atomic oxygen (Craig and Gille, 1969). The IR curve represents the heat loss (or gain) due to infrared radiation, based on the calculations of Kuhn and London (1969). Figure 2 is similar to Figure 1 except the IR curve is based on calculations by Kondratiev et al. (1966) rather than Kuhn and London; it shows considerably lower infrared losses near the mesopause and shows a heat input between 65 and 80 km.

Figure 3 shows a family of curves for different latitudes at the solstice; the individual curves are assembled in the same fashion as the total curve in Figure 1. Also shown is an average curve; this is a weighted average over the earth of the other curves and represents the world-wide average of excess heat input above given altitudes over

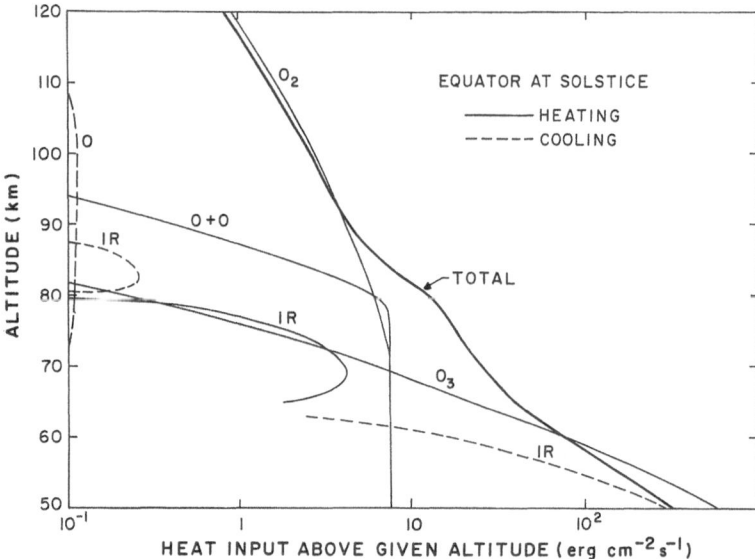

Fig. 2.   Similar to Figure 1 but using the infrared heat losses calculated by Kondratiev *et al.* (1966).

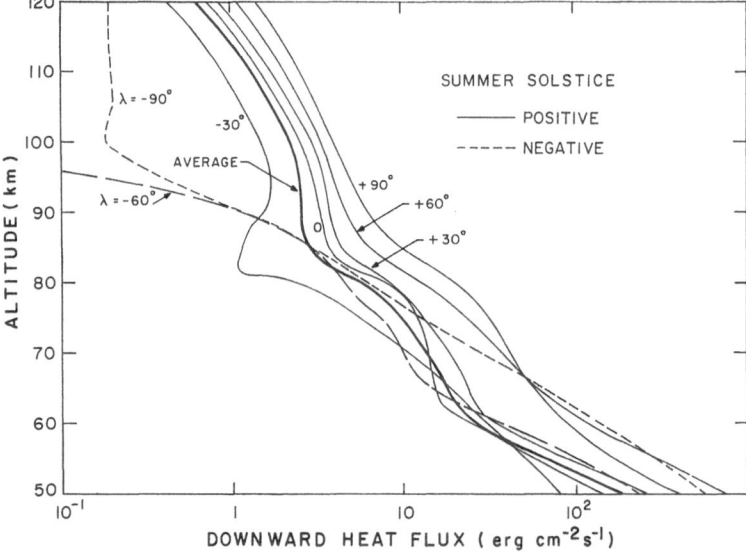

Fig. 3.   Net heat input into the upper atmosphere above given altitudes during the solstice, according to Johnson and Gottlieb (1970). The curve labeled O is for the equator and is the total curve shown in Figure 1. Curves are shown for various latitudes, with summer latitudes shown as positive.

losses other than by eddy conduction. This excess heat input must be compensated by a downward eddy heat flux that transfers the heat to lower altitudes (below 65 km) where it can be lost by infrared radiation to space.

Figure 4 indicates the average profile of eddy mixing required to transfer downward the excess heat deposited in the atmosphere above each altitude. The required coefficients vary from about 10^7 cm^2 s^{-1} at 110 km down to 10^5 near 60 km. Above 110 km, molecular conduction overwhelms eddy conduction and no means exist of estimating rates of eddy mixing from the heat balance. However, it must be anticipated

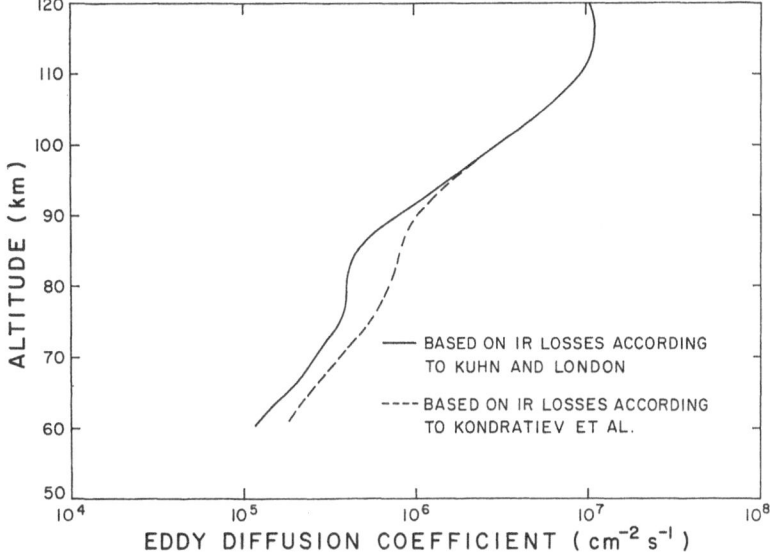

Fig. 4. The profile of eddy mixing coefficient required as a worldwide average to transport downward the excess heat input above each altitude indicated by the average curve in Figure 3, according to Johnson and Gottlieb (1970). The solid curve shows the results based on the study of infrared losses by Kuhn and London (1969) whereas the dashed curve is based on the study by Kondratiev *et al.* (1966).

that the eddy diffusion coefficient will start decreasing with altitude not far above 110 km. The reason for this is that eddy lifetimes will be shortened by molecular diffusive dissipation of eddies; the exponentially increasing rate of molecular diffusion, which becomes larger than the eddy coefficient near 110 km, must make eddy transfer ineffective within a scale height or two above the altitude where the two are equal (Johnson and Gottlieb, 1970). Keneshea and Zimmerman (1970) have also noted the expected termination of turbulence due to molecular diffusion but claim an effective cessation of turbulence at that altitude where "the kinematic viscosity is of the order of the turbulent viscosity" rather than a scale height or so above it, as concluded by Johnson and Gottlieb.

3. Composition Transport in the Vertical

Colegrove *et al.* (1965), following a method of calculation equivalent to that developed by Lettau (1951) for the combined transport effects of molecular and eddy motion, calculated profiles of atomic and molecular oxygen concentration to be expected in the lower thermosphere for various assumed rates of eddy mixing where the mixing coefficient was assumed to be constant with altitude. They calculated rates of photo-dissociation and recombination of oxygen and applied the continuity condition to determine the transport of atomic oxygen downward and molecular oxygen upward. The photodissociation rates were not varied continuously throughout the day, but instead a 45° zenith angle was assumed to apply for 12 hours a day; this gave a good approximation to the total dissociation per day and to its height distribution. Their results are shown in Figure 5 for three different rates of eddy mixing (Colegrove *et al.*, 1966). Note that with the highest rate of eddy mixing, the atomic oxygen concentrations are lowest because of the rapid eddy transport downward from the region of photodissociation to the region of recombination. This also maintains the highest concentration of molecular oxygen at the upper levels due to the more rapid replenishment of the photodissociation losses; however, the molecular oxygen distribution is not as sensitive to changes in the rate of eddy mixing as is the atomic oxygen distribution. The middle curves (solid) are in best agreement with observed atomic oxygen concentrations, corresponding to a value of unity at 120 km for the atomic-to-molecular oxygen ratio.

While Colegrove *et al.* assumed an equilibrium situation and showed that time constants were large enough to validate this concept, Shimazaki (1967) approached the same problem from a different viewpoint, which he characterized as time de-

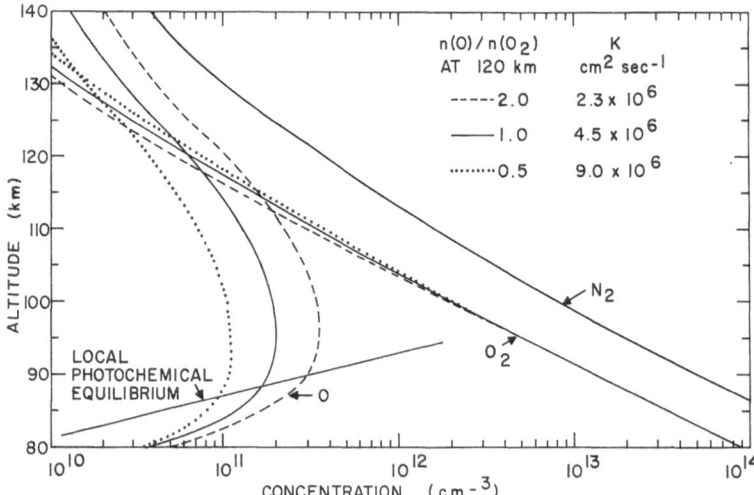

Fig. 5. Concentrations of the major atmospheric constituents taking into account photodissociation, recombination, and vertical transport for three values of the eddy diffusion coefficient K, according to Colegrove *et al.* (1966).

pendent. However the time dependence that he introduced is not the diurnal effect but rather the rate at which equilibrium conditions are approached, or the time required to produce equilibrium distributions of the sort calculated by Colegrove *et al.* He assumed normal solar incidence, 24 hours a day. Figure 6 shows a series of calculated atomic oxygen distributions for various times after starting with an arbitrary distribution; these indicate a time constant of a few days for approaching equilibrium, and this would become a week or two if the solar radiation were made

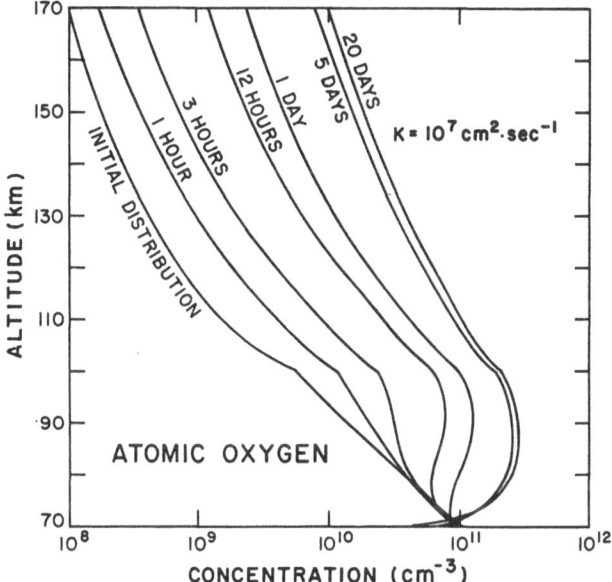

Fig. 6. Concentrations of atomic oxygen after various periods of time, taking into account photodissociation, recombination, and vertical transport, according to Shimizaki (1967).

to vary diurnally. The break in the curves at 100 km is due to a discontinuity in the assumed temperature distribution and has no other physical significance. Figure 7 shows the equilibrium atomic and molecular oxygen distributions calculated by Shimazaki, and these may be compared with those of Colegrove *et al.* shown in Figure 5. Both calculations show a broad maximum near 90 km. Shimazaki's calculations were based upon an eddy diffusion coefficient of 10^7 cm^2 s^{-1} and should be most nearly comparable to Colegrove *et al.*'s calculations utilizing a coefficient of 9×10^6 cm^2 s^{-1}. The greater atomic oxygen concentrations calculated by Shimazaki are due to his assumption of continuous, normal-incidence sunlight, giving rise to about three times more dissociating events per day than in the calculations of Colegrove *et al.*

Shimazaki's distribution does not fall off as rapidly with decreasing altitude near 80 km because he considered the recombination process, $O + O + M \rightarrow O_2 + M$, whereas $O + O_2 + M \rightarrow O_3 + M$ was included in Colegrove *et al.*'s calculations. Otherwise the two calculations appear to be in excellent agreement. Shimazaki's selection of an eddy mixing coefficient of 10^7 cm^2 s^{-1} to fit the observational data on atomic and molecular oxygen is too high because of his assumption of continuous solar radiation; Colegrove et al.'s value of 4.5×10^6 cm^2 s^{-1} to fit the same data is more realistic.

It is interesting to examine the noble gas concentration profiles calculated by

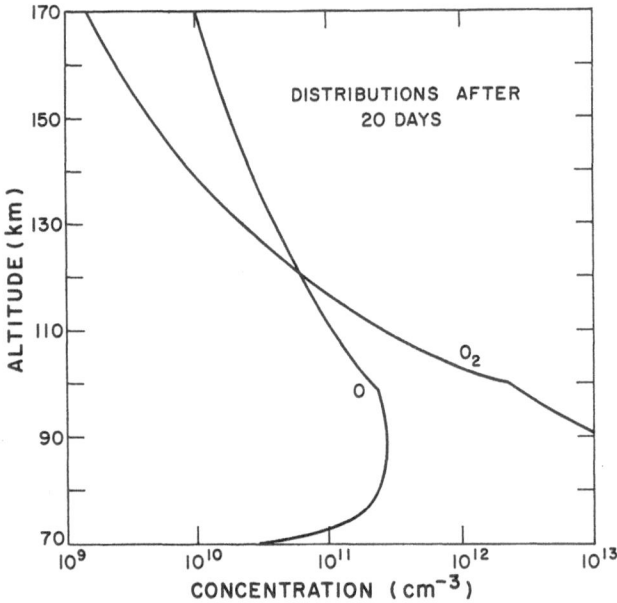

Fig. 7. Equilibrium concentrations of atomic and molecular oxygen, taking into account photo-dissociation, recombination, and vertical transport, according to Shimazaki (1967).

Colegrove et al. (1966). Figure 8 shows profiles calculated for three different rates of eddy mixing, assumed to be uniform in time and altitude. Helium is much lighter than the average atmospheric gas and, under action of molecular diffusion, tries to flow out of the lower atmosphere and up into the thermosphere. Eddy mixing acts as a pump to transport helium from above the turbopause to below the turbopause. The equilibrium concentration curve shown for helium in Figure 8 involves an upward transport by molecular diffusion that just matches the downward transport by eddy diffusion. Argon, heavier than the average atmospheric gas, tries under action of molecular diffusion to sink into the lower atmosphere, but eddy mixing acts as a pump to maintain its concentration in the upper atmosphere higher than it would otherwise be. This it does well up to an altitude of about 100 km, but higher up it loses out to molecular diffusion, which maintains a diffusion equilibrium distribution above about 130 km. Between 100 and 130 km is a transition region, which is neither well mixed nor in diffusion equilibrium.

It is reasonable to define the turbopause as that altitude at which the molecular diffusion coefficient equals the eddy diffusion coefficient. Since the molecular diffusion coefficient depends upon the constituent in question, this concept requires that different turbopauses be defined for different species. This is illustrated in Figure 8 for three assumed rates of eddy mixing; the three helium turbopauses are lower than the three argon turbopauses because the molecular diffusion coefficient for helium is larger than for argon, other conditions being equal, and it is necessary to go to a higher altitude for argon than for helium to attain a molecular diffusion coefficient as large as the accepted value for the eddy diffusion coefficient.

Hodges (1970) has drawn attention to the fact that internal gravity waves, even in the absence of turbulence, act to transport lighter gases downward and heavier gases upward. Thus this process acts in the same sense as turbulence and supplements it. Hodges concludes that gravity wave transport is a factor in decreasing the density scale height for helium from the values calculated on the basis of diffusion equilibrium above the turbopause.

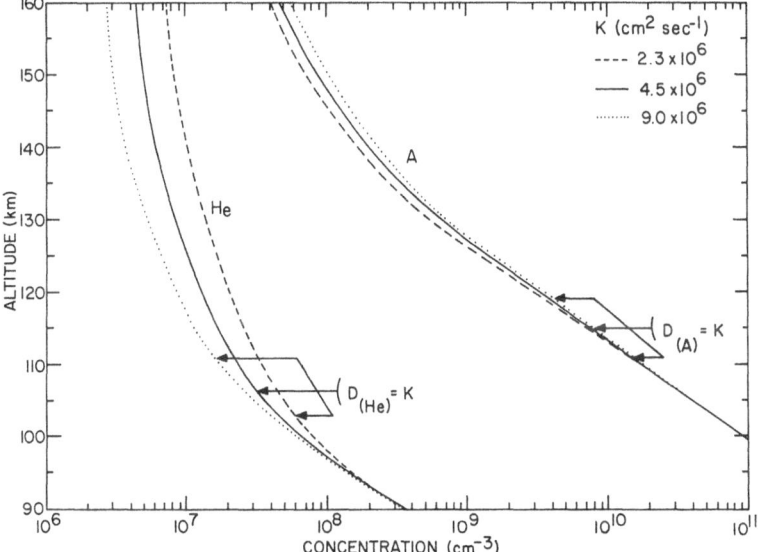

Fig. 8. Concentrations of argon and helium for three values of the eddy diffusion coefficient K, according to Colegrove *et al.* (1966). Arrows indicate the altitude at which the molecular diffusion coefficient equals the eddy diffusion in each case.

4. Horizontal Transport

Figure 3 shows a non-uniform heat input over the earth above various altitudes. However, except for the diurnal bulge, the pattern of upper atmospheric temperature is amazingly uniform. The diurnal minimum is cooler than the winter polar region. This suggests that something is acting to even out the non-uniformity of heat input, and the only likely mechanism is wind. Johnson and Gottlieb (1970) have calculated the average meridional wind required to accomplish this. The method used was to calculate the vertical motion and compressional heating or cooling required to compensate for the non-uniformity in heat input as a function of latitude and altitude shown in Figure 3. The average vertical motion required at the solstice is shown in Figure 9 for several latitudes; it is upward over the equator and the entire summer hemisphere and downward in the winter hemisphere except within perhaps 15° of the equator. Figure 10 shows the associated horizontal wind required at low latitude

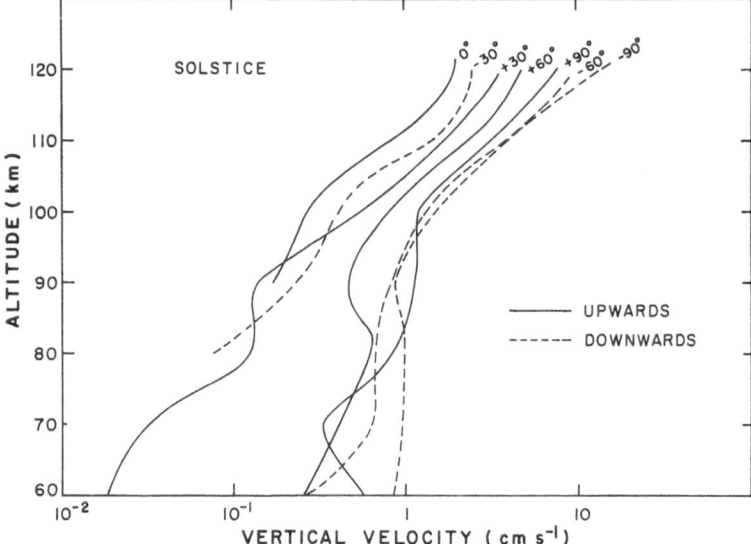

Fig. 9. Vertical velocities calculated for various latitudes to compensate for unsymmetrical solar heating input into the upper atmosphere at the solstice, according to Johnson and Gottlieb (1970). Positive labels apply to the summer hemisphere, negative to the winter.

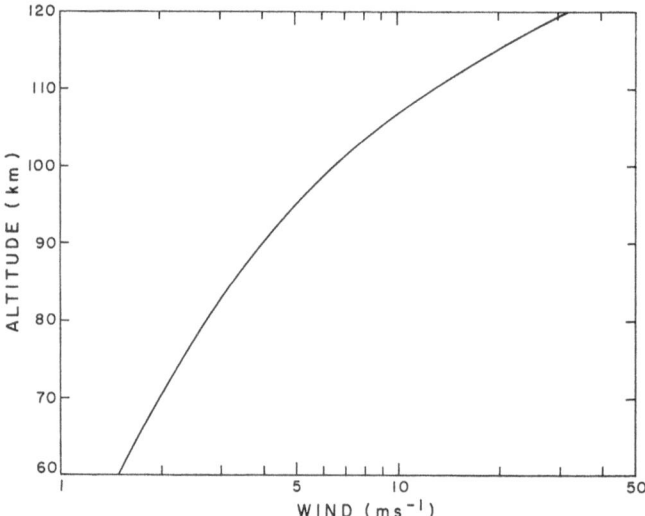

Fig. 10. Average meridional winds required to provide the vertical velocities shown in Figure 9, according to Johnson and Gottlieb (1970). The values shown are equatorial, and the latitudinal variation has been assumed to be according to cos λ.

to maintain continuity where the wind is assumed to vary with latitude λ as $\cos\lambda$. This then is the meridional wind system required at solstice to maintain the warmth of the winter polar region at the expense of solar heat input into the summer hemisphere.

The meridional wind system indicated in Figure 10 provides the most probable explanation for the winter helium bulge, a build up in helium concentration at high altitudes over the winter polar region (Keating and Prior, 1969). The horizontal transport of a light minor atmospheric constituent is enhanced relative to the major atmospheric constituents in proportion to the scale height (Johnson and Gottlieb, 1970).

Kellogg (1961) has drawn attention to the possible importance of recombination of atomic oxygen as a heat source in the mesosphere over the winter polar region, associated with downward currents there. However, the study of Johnson and Gottlieb (1970) indicates that the heat transfer associated with atomic oxygen transported in this manner is relatively unimportant, amounting to only about 20% of the heat release due to oxygen recombination associated with downward transport by small scale eddies.

Kohl and King (1967) have calculated the diurnal winds to be expected in the upper thermosphere for an equinoctial situation. They used the density distribution deduced by Jacchia (1964) from satellite drag data, simplified to an idealized symmetrical distribution of pressure. They took earth rotation, ion drag, and viscosity into account and calculated the wind patterns shown in Figures 11 and 12 for an altitude of 300 km. The distributions are quite symmetric, generally blowing from the diurnal density

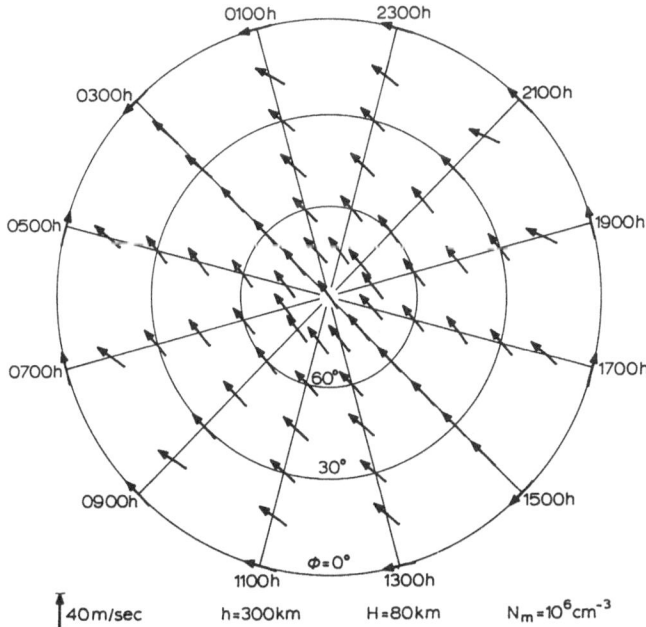

Fig. 11. The atmospheric wind system in the northern hemisphere calculated for an altitude of 300 km when the peak electron density is 10^6 cm^{-3}, according to Kohl and King (1967).

maximum to the minimum. It is reasonable to suggest on the basis of the study of
Johnson and Gottlieb that there should be a net inflow into both polar regions to
make up for heat input deficits there at the equinox. These are not evident in the
wind patterns calculated by Kohl and King, but they presumably would appear if
accurate data were available for the pressure gradient distributions over the polar
regions. They would appear as stronger winds into the polar region from the diurnal
maximum than out of the polar region into the diurnal minimum, thus producing
a net inflow into the polar region.

Rishbeth *et al.* (1969) have carried the calculation of diurnal wind systems a step
farther by determining the horizontal divergence in the calculated wind patterns.
In addition, they discriminate between the diurnal expansion of the atmosphere,
which increases pressures at all altitudes in the upper atmosphere but does not increase
the pressure experienced by any air particle that moves with the bulk atmosphere in
its expansion, and horizontal divergence, which reduces the pressure experienced by
such air particles. In the case of the approach of the diurnal pressure maximum,
the atmosphere expands and the pressure increases at all altitudes but this in itself
does not cause individual air particles to experience a pressure change. However,
it causes horizontal divergence in the wind field which leads to pressure decreases.
However, the pressure decrease associated with the divergence cannot be as large
as the diurnal increase, otherwise it would overcompensate the effect that produces
the divergence in the first place. The situation near the diurnal minimum is similar

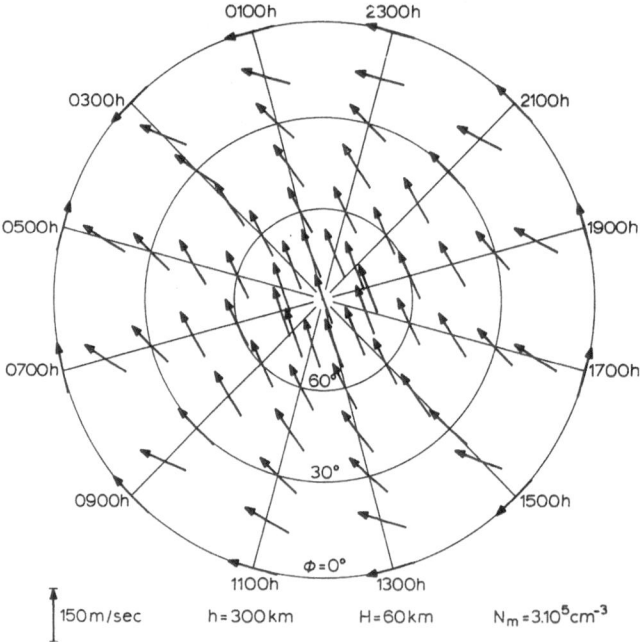

Fig. 12.   The atmospheric wind system in the northern hemisphere calculated for an altitude of
300 km when the peak electron density is 3×10^5 cm^{-3}, according to Kohl and King (1967).

but opposite in sense. Figure 13 shows the pattern of vertical velocity W_D due to divergence and W_B due to thermal expansion; also shown are the concentrations of atmospheric particles through the day at sunspot minimum and maximum at 300 km. Figure 14 shows the vertical pattern of vertical velocity due to divergence; also shown is the associated flux of particles, which decreases with altitude because of the decrease in atmospheric density with altitude. The results become unreliable at the lowest altitudes shown because zero wind was assumed at 120 km in the calculations. The results apply for 45° latitude at equinox.

Volland (1969) has considered the effect of diurnal tidal waves near the equator

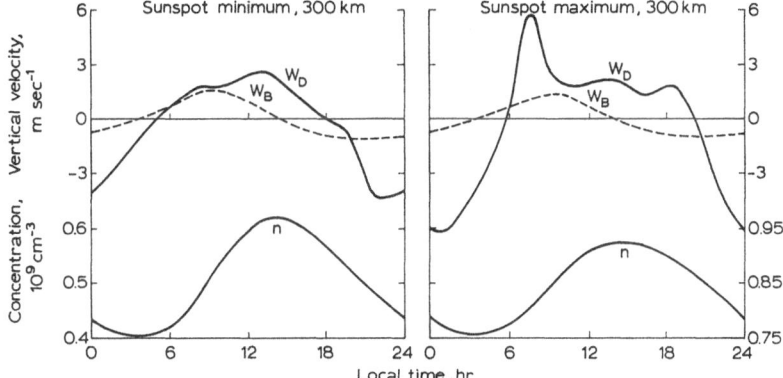

Fig. 13. Local time variation of the divergence velocity W_D and the barometric velocity W_B at 300 km (both positive upwards) according to Rishbeth et al. (1969). The variation of atomic concentration at 300 km is shown below.

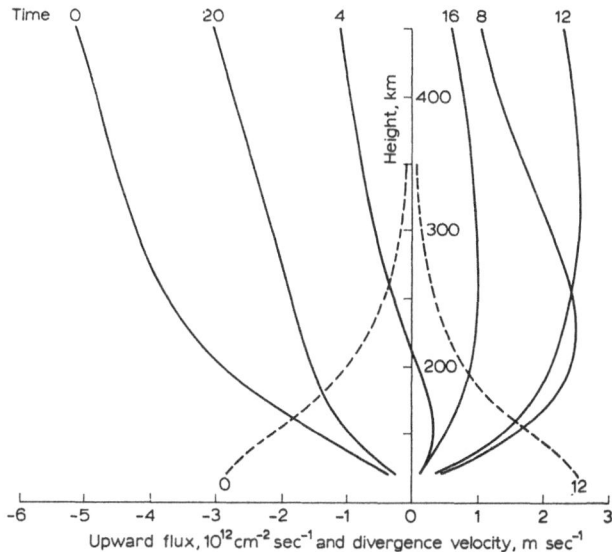

Fig. 14. Altitude variation of the vertical flux (----) at midnight (0) and noon (12) and of the divergence velocity W_D at six different local times (———), according to Rishbeth et al. (1969).

and concludes that they may dominate over other circulations up to 250 km altitude. By proper choice of tidal wave amplitude and phase, he claims to reproduce the observed phase of the diurnal density variation as observed by satellites. This is another way of introducing horizontal winds into the problem of upper atmosphere density variations. It is in a sense arbitrary, and it ignores the effects of north-south transport. However, there can be no doubt that tidal variations must be included in any full and correct treatment of upper atmospheric variations.

In conclusion, it is clear that major wind systems play an important role in limiting the magnitude of both diurnal and seasonal variations in atmospheric density. However, studies of the sort discussed here will require considerable extension before all the major effects associated with these wind systems are identified. Some of the important factors requiring better explanation are the seasonal and possibly diurnal composition changes and the phase of the diurnal variations.

Acknowledgement

This research was supported by the National Aeronautics and Space Administration under grant NGL 44-004-026.

References

Colegrove, F. D., Hanson, W. B., and Johnson, F. S.: 1965, *J. Geophys. Res.* **70**, 4931.
Colegrove, F. D., Johnson, F. S., and Hanson, W. B.: 1966, *J. Geophys. Res.* **71**, 2227.
Craig, R. A. and Gille, J. C.: 1969, *J. Atmospheric Sci.* **26**, 205.
Emden, R.: 1913, 'Über Strahlungsgleichgewicht und atmosphärische Strahlung', *Sitzber. K. Bayer, Wiss. Gesellsch. München, Math.-Phys. Kl.*, **43**, 55–142.
Epstein, P. S.: 1932, *Gerlands Beitr.* **35**, 153.
Gowan, E. H.: 1936, *Q. J. Roy. Meteor. Soc. Suppl.* **62**, 34.
Hesstvedt, E.: 1968, *Geofys. Publikasjoner, Geophys. Norv.* **27**, 1.
Hodges, R. R., Jr.: 1970, 'Vertical Transport of Minor Constituents in the Lower Thermosphere by Nonlinear Processes of Gravity Waves', submitted to *J. Geophys. Res.*
Jacchia, L. G.: 1961, *Space Res.* **2**, *Proc. of the Second International Space Science Symposium, Florence* (ed. by H. C. van de Hulst, C. de Jager, and A. F. Moore), North-Holland Publishing Co., p. 747.
Jacchia, L. G.: 1964, Smithsonian Astrophysical Observatory Special Report No. 150, 32 pp.
Jeans, J. H.: 1916, *Dynamical Theory of Gases*, 2nd ed., Cambridge.
Johnson, F. S. and Gottlieb, B.: 1970, Eddy Mixing and Circulation at Ionospheric Levels', submitted to *Planetary Space Sci.*
Johnson, F. S. and Wilkins, E. M.: 1965, *J. Geophys. Res.* **70**, 1281; Correction, *J. Geophys. Res.* **70**, 4063.
Keating, G. M. and Prior, E. J.: 1969, *Space Res.* **8**, 982.
Kellogg, W. W.: 1961, *J. Meteorol.* **18**, 373.
Keneshea, T. J. and Zimmerman, S. P.: 1970, 'The Effect of Mixing Upon Atomic and Molecular Oxygen in the 70–170 km Region of the Atmosphere', submitted to *J. Atmospheric Sci.*
Kohl, H. and King, J. W.: 1967, *J. Atmospheric Terrest. Phys.* **29**, 1045.
Kondratiev, K. Ya., Badinov, I. Y., Gaevskaya, G. N., Nikolsky, G. A., and Shved, G. M.: 1966, 'Radiative Factors in the Heage Regime and Dynamics of the Atmospheric Layers', in *Problems of Atmospheric Circulation* (ed. by R. V. Garcia and T. F. Malone), Macmillan and Co., Ltd., London.
Kuhn, W. R. and London, J.: 1969, *J. Atmospheric Sci.* **26**, 189.

Lettau, H.: 1951, *Compendium of Meteorology* (ed. by T. F. Malone), American Meteorological Society, pp. 320–333.

Nicolet, M.: 1954, *J. Atmospheric Terrest. Phys.* **5**, 132.

Nicolet, M.: 1961, *J. Geophys. Res.* **66**, 2263.

Rishbeth, H., Moffett, R. J., and Bailey, G. J.: 1969, *J. Atmospheric Terrest. Phys.* **31**, 1035.

Shimazaki, T.: 1967, *J. Atmospheric Terrest. Phys.* **29**, 723.

Spitzer, L.: 1949, *The Atmospheres of the Earth and Planets* (ed. by G. P. Kuiper), University of Chicago Press, pp. 211–247.

Volland, H.: 1969, *J. Atmospheric Terrest. Phys.* **17**, 1581.

THE EXOSPHERE AND GEOCORONA

P. MANGE

E. O. Hulburt Center for Space Research, U. S. Naval Research Laboratory,
Washington, D.C., U.S.A.

Abstract. This paper reviews the observational evidence on the hydrogen geocorona including its density, vertical structure, and the nature of its diurnal and other solar-related variability. The relationship of observationally-derived models of the exosphere and geocorona to the thermosphere is indicated. The hydrodynamic and kinetic descriptions of the origin of the polar wind, the outward supersonic ion flow over the poles, are briefly reviewed. Observations of vertical ion distributions, of the supersonic flow of ionized atomic hydrogen, and of the polar depletion of atomic hydrogen are cited as evidence for the polar wind. These results are interpreted as clues to the geographical extent of the polar wind and its correlation with the plasmapause.

1. Introduction

The earth's exosphere is that region in which neutral atomic collisions are infrequent, and whose density distribution is therefore directly coupled to constituent distributions in the thermosphere just below, where collisions are frequent. As is well known, the distribution of the heavier atoms, which follow ballistic orbits in the exosphere, is readily calculated. To describe the behavior of helium and hydrogen is more difficult, for they are the only species which undergo thermal evaporation from the top of the atmosphere. Nevertheless, the outward flow of escaping neutral helium is so small, when compared with its concentration in the atmosphere, that the flow may be neglected. Thus, the exospheric winter helium bulge of Keating and Prior (1968) is intimately tied to its circulation at lower levels, as Hodges and Johnson (1968) have shown. We neglect the question here except for incidental observational evidence to be presented further on.

2. The Outward Flow of Atomic Hydrogen

On the other hand, atomic hydrogen, the principal escaping component, arises from mesospheric dissociation of methane and water vapor, as originally discussed by Bates and Nicolet (1950), and is carried upward by mixing, but with concentration falling as altitude increases. When a certain level in the region above 100 km is reached maximum molecular diffusion acts to support the flow, which moves upward against gravity through the thermosphere and out to the exosphere.

The theory of the transition flow across the exobase has been unsatisfactory. In a review by Chamberlain and Campbell (1967) earlier analytic studies were supplemented by their own Monte Carlo calculations which were then compared with the Monte Carlo analyses of two other investigating teams. No two of the three numerical treatments were consistent, and the question has remained unresolved until

Dyer (ed.), Solar-Terrestrial Physics/1970: Part IV, 68–86. All Rights Reserved.
Copyright © 1972 by D. Reidel Publishing Company.

the recent publication of Brinkman (1970). He verifies the findings of Chamberlain and Campbell. The ratios of calculated escape fluxes to those of the classical Jeans escape theory are typically found to be 0.70–0.75 and 0.97–0.99 for hydrogen and helium, respectively.

3. Geocoronal Hydrogen and the Lyman-α Glow

The fundamental problem of the exosphere has been to verify and refine theoretical models of the hydrogen distribution using observational evidence. As shown in Figure 1 (Meier and Mange, 1970) Meier has matched models due to Kockarts and Nicolet (1963) for the thermosphere with those of Chamberlain (1963) for the exosphere. The latter take account of the presence (a) of no hydrogen in satellite exospheric orbits (parameter $R_{sc} = 1.0\ R_c$), (b) of some hydrogen in satellite orbits ($R_{sc} = 2.5\ R_c$) or (c) of maximum possible orbiting hydrogen ($R_{sc} = \infty$). The models are adjusted to a reference concentration of $3 \times 10^7\ cm^{-3}$ at 100 km, which corresponds to $6 \times 10^4\ cm^{-3}$ at 650 km for a temperature of 1100 K. This absolute determination is made by observing the hydrogen Lyman-α radiation at 1216 Å in the sky from satellites.

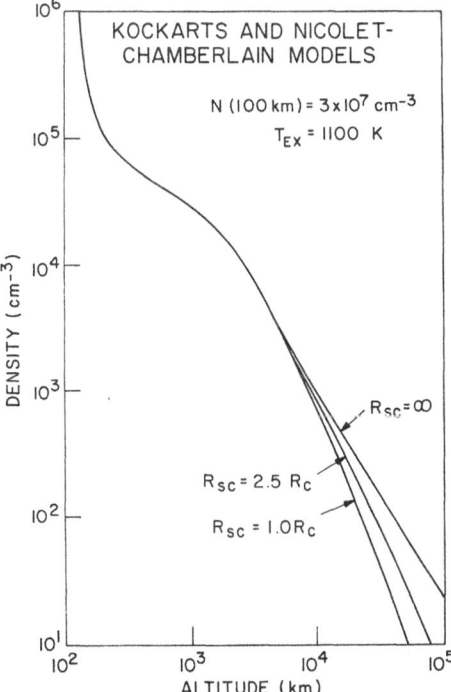

Fig. 1. Hydrogen concentration as a function of altitude for Kockarts and Nicolet-Chamberlain models. The composite model for $T_{EX} = 1100\,K$ has been adjusted in absolute concentration to fit Lyman-α data taken from the OGO-4 satellite. The three branches at high altitude correspond to models with no, with some, or with maximum hydrogen in satellite orbits.
(After Meier and Mange, 1970.)

The observed glow originates in sunlight scattered around the earth by the geocoronal hydrogen, as first suggested by Johnson and Fish (1960). Second order radiative transfer theory developed and applied by Thomas (1963), Donahue (1967), and by Meier (1969) yields the concentration.

Recently Thomas *et al.* (1970) have independently interpreted results from satellite OGO-6 which yield concentrations of order 10^5 hydrogen atoms cm^{-3} in basic agreement with the above models. There are contrasting reports of neutral mass spectrometer observations from rockets (Reber *et al.* 1968; Bjuro *et al.* 1969) which show as much as two orders of magnitude more hydrogen. These findings are wholly inconsistent with the optical results and cannot be accepted. We note, furthermore, that early altitude-dependent observations of scattering from the center of the solar Lyman-α line, which were reported by Tousey (Mange *et al.* 1960), yielded rough correspondence with the hydrogen densities in the present models. Indeed, those data were used by Bates and Patterson (1961a, b) as criteria for the selection of models in their early comprehensive study of hydrogen in the exosphere. Meier's analysis of data from the OGO-4 satellite indicates support for Chamberlain's preferred distribution of partially populated satellite orbits corresponding to $R_{sc} = 2.5\ R_c$. But we shall review some contrasting evidence from a high altitude satellite below.

4. Variability of Geocoronal Hydrogen

The evaporation factor for loss of hydrogen to space is strongly dependent on temperature. Since by continuity the constant supply of hydrogen from below must be matched by the flow to space, the equilibrium concentration of outflowing hydrogen is reduced in the exosphere when the temperature rises (and the evaporation coefficient is high) as in daytime. Conversely, the concentration rises at night. From early intensity calculations, Donahue and Thomas (1963) had estimated a diurnal concentration variation of 5 to 1. However, Hanson and Patterson (1963) estimated the lateral flow which might occur in an atmosphere of cylindrical geometry with unequal heating. They found that the diurnal variation would be reduced by lateral flow to no more than a ratio of 2 to 1. Further extension of the theory by Patterson (1966) to incorporate the effects of lateral flow in full time-dependent solutions led to diurnal concentration ratios which were less than two for low temperature models, but greater than two for high temperatures. Refinement of the theory by McAfee (1967) with use of spherical geometry resulted in a prediction of at least a 2-to-1 diurnal variation.

Figure 2 depicts observations reported by Meier and Mange (1970) in which the intensity in kilorayleighs is plotted as a function of solar zenith angle. The three curves correspond to theoretical intensities predicted from hydrogen models, which were constructed to vary through night and day around the earth by factors of 1, 2 and 4 relative to the concentration at the base of the exosphere for solar zenith angle of 90°. (Thus, "1" corresponds to complete spherical symmetry or uniform distribution of the hydrogen over day and night.) From these data the upper limit of variability is judged to be 4 and could conceivably be much less.

Figure 2 further illustrates the notably good fit of theory to data, referred to above, for the model $R_{sc}=2.5\,R_c$, in which the satellite orbits of atomic hydrogen in the exosphere are only partially populated. At much higher altitudes the intensity of solar Lyman-α scattered back by atomic hydrogen from beyond 5 earth radii was measured from the OGO-3 satellite (Mange and Meier, 1970), and showed marked changes from one orbit to the next. For example, the intensities observed on June 15–16,

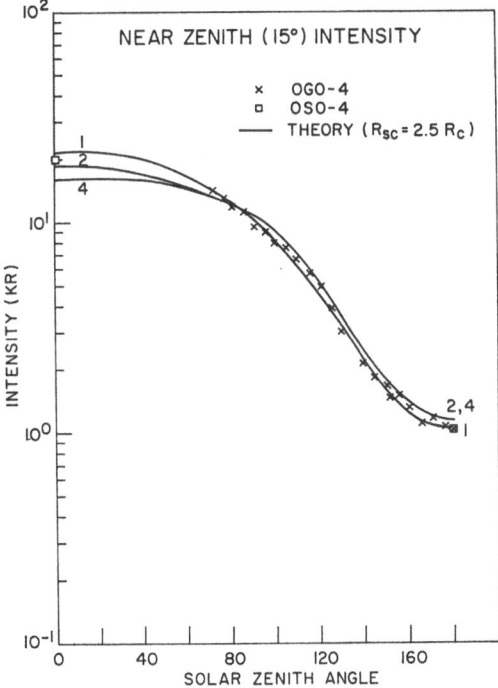

Fig. 2.   Near zenith experimental and theoretical intensities of Lyman-α as a function of solar zenith angle. The theoretical profiles correspond to models with diurnal variations of 1, 2 and 4 in concentration at the base of the exosphere. (After Meier and Mange, 1970.)

1966 were an excellent fit to the $R_{sc}=2.5\,R_c$ model, and would suggest concentrations of the order of 25 atoms cm^{-3} at 50000 km. However, as seen in Figure 3, the distribution in the 35000–80000 km range on June 17–18, 1966 implied the absence of hydrogen in satellite orbits. Rather than speculate as to how this might arise we await a more definitive description of the phenomenon which should be contained in OSO-4 data still to be analyzed.

In addition to the questions of geocoronal concentration, satellite orbit population, and lateral flow we may ask whether there is any apparent geocoronal dependence on the solar cycle. For this purpose we use data from a Lyman-α glow experiment flown to 900 km during solar minimum in 1964 (Meier *et al.*, 1970). In Table I is presented the appropriate exospheric model based on the low 750° temperature inferred from satellite drag measurements. Intensities computed from this model, designated A, are

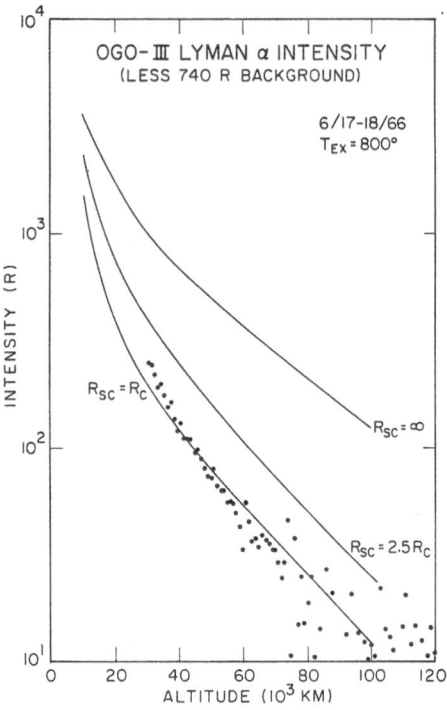

Fig. 3. Lyman-α intensity from satellite OGO-3 as a function of altitude with background sub-tracted. The theoretical intensities correspond to the adopted Chamberlain profiles. (After Mange and Meier, 1970.)

TABLE I

Hydrogen concentration model (A) for $T_{EX} = 750°$

Altitude	Concentration
100 km	$3 \times 10^7 \text{ cm}^{-3}$
650 km	$5 \times 10^5 \text{ cm}^{-3}$
10 000 km	$2.5 \times 10^3 \text{ cm}^{-3}$
50 000 km	$3.6 \times 10^1 \text{ cm}^{-3}$

compared in Figure 4 with data obtained when looking upward (at 38° from zenith). Intensities for two other models, B and C, reduced in density from A by a factor of 3 and 6, respectively, are also shown. Plainly, high density model A does not fit, and, after weighing the possible effect of solar Lyman-α emission, and of extraterrestrial background, we conclude that there is evidence for a variation in hydrogen production over the solar cycle of perhaps a factor of two.

Incidentally, we may take note of a severe problem in interpreting sounding rocket observations of the geocorona. For in Figure 5 we again see plots of intensity from above (in the upper curve), and also from below (in the lower curve). They show that the intensities during ascent, at the left, are severely depressed relative to those of descent, on the right, presumably because they are seen through an absorbing layer of gas carried along with the rocket as it rises. Our interpretive data were based on the descent curve.

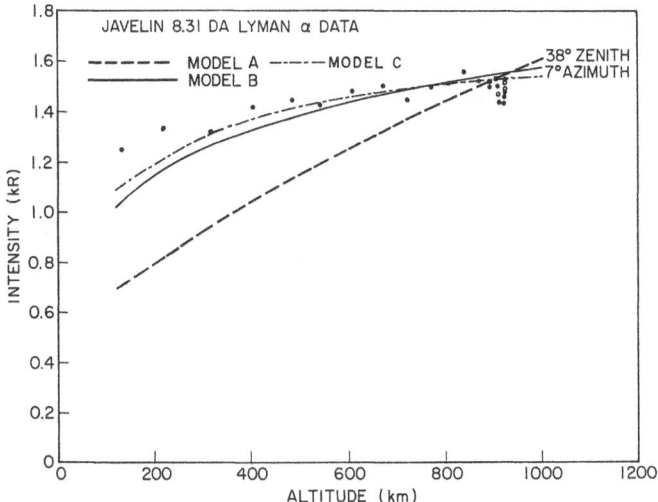

Fig. 4.   Experimental and theoretical Lyman-α emission rates as a function of altitude from a Javelin rocket (NASA 8.31DA) flown January 17, 1964 near midnight local time from Wallops Island, Va. The data correspond to direction-of-look closest to local zenith. The small circles show the data during descent after correction for outgassing. No correction was made for extraterrestrial background. (After Meier *et al.*, 1970.)

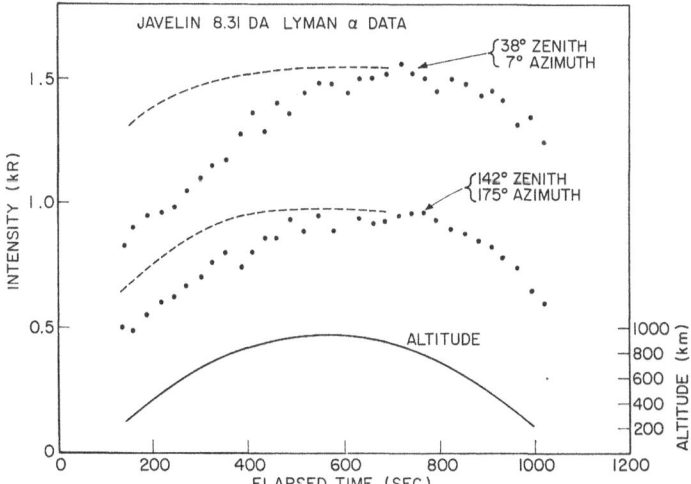

Fig. 5.   Lyman-α intensity from Javelin rocket (NASA 8.31DA) as a function of elapsed time from launch for directions corresponding to closest approach to local zenith and nadir. Rocket altitude is also shown. The dashed lines show a curve corrected for absorption caused by outgassing. (After Meier *et al.*, 1970.)

5. Observation of Other Emissions from the Exosphere

Of course Lyman-α radiation is not the only optical line by which one can deduce
the hydrogen distribution in the atmosphere. A paper in this symposium by Tinsley
and Meier (1972) discusses deductions to be made from Balmer-α observations of the
celestial sphere. Recently, Weller *et al.* (1970) have flown a photometer at night which
was sensitive in the range 740–1050 Å which includes the Lyman-β line of hydrogen
at 1026 Å. In Figure 6 we show their projection of intensity contours in this line in

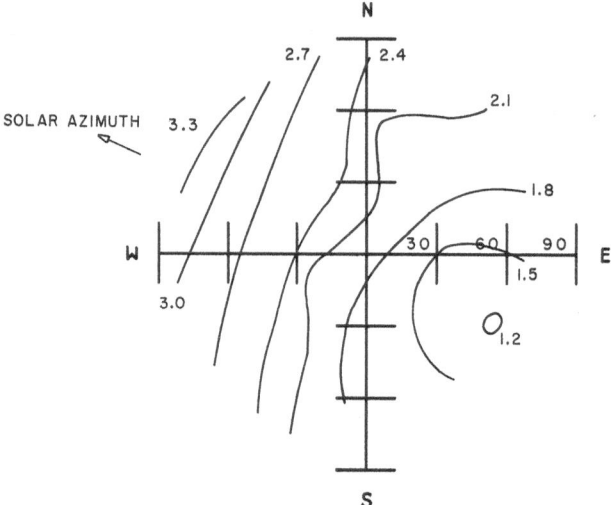

Fig. 6.   Contour plot of 740–1050 Å photometer data measured in thousand counts/second at night
(Solar zenith angle of 134°). Local zenith angle is measured radially from center. (Weller *et al.*, 1970.)

thousands of counts/second over the celestial sphere. The arrow indicates the azimuthal
direction of the sun which is 44° below the horizon. The essential symmetry of the
contours about the solar direction, and the intensity increase toward it, is apparent.
On this flight other photometers were carried which by subtraction could in principle
provide very rough measure of the resonance line of neutral helium at 584 Å in the
atmosphere. If the observed source was indeed 584 Å its zenithal intensity would not
be greater than two rayleighs. However, the precision in the derived intensities was
low; and it was concluded that there was no definite evidence of 584 Å radiation in
the residual signal. Emission in the 304 Å line itself was unambiguously charted and
was clearly not symmetric about the solar azimuth, but increased in a direction much
to the south of the solar direction. Thus, if the pattern is produced by resonant scattering
of sunlight, there is a strong asymmetry in the distribution of helium ions about the
earth-sun line.

6. The Ion Exosphere and the Polar Wind

Now we turn to discuss the ion exospheric behavior in the polar regions, and especially
the polar wind (a term applied by Axford (1968) in recognition of its supersonic

character, analogous to the solar wind). Much observational evidence which bears on the polar wind is presented at this symposium in sessions concerned with the polar ionospheric structure, whistlers, and the plasmapause. As introduction to the topic we present new ion composition data obtained by Brinton and Grebowsky (1970). In Figure 7 their simplified symmetric diagram of the plasmasphere is shown. Data were taken in the blacked-in regions shown within the permanently closed field lines

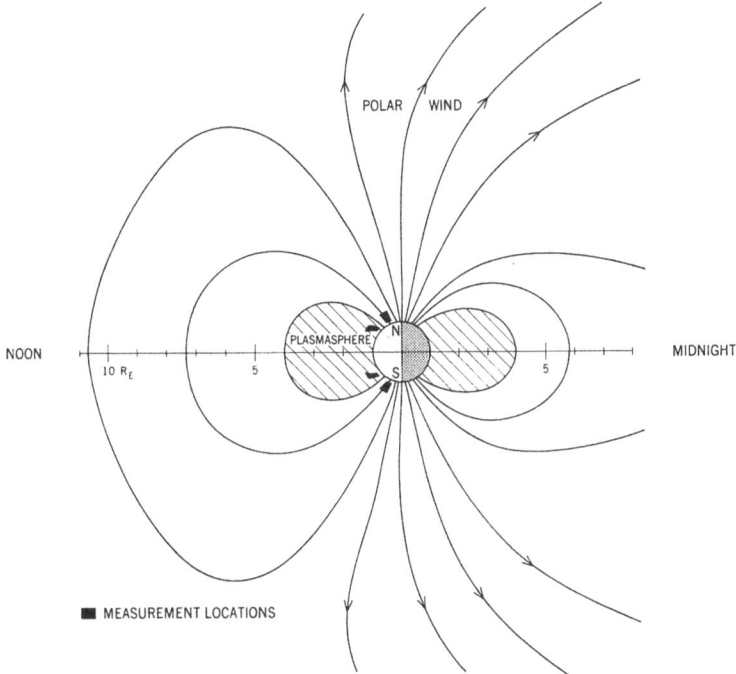

Fig. 7. Symmetric diagram of the plasmasphere. (Brinton and Grebowsky, 1970.)

of the plasmasphere, and also within the region of convecting field lines. These open and close as the earth rotates, in accordance with the developed theory of Axford and Hincs (1961) and Nishida (1966). Within the plasmasphere the data of Brinton and Grebowsky in Figure 8 show that O^+ and H^+ follow the altitude distributions predicted from the charge separation theory of Pannekoek and Rosseland (cf. Lemaire and Scherer, 1970a). Thus, in the region where O^+ is predominant the tendency of electrons to escape creates a strong polarization field with O^+ that exerts a force equal to half the gravitational force on the O^+ ion. The very light and less abundant H^+ ions are thus buoyed up and increase with altitude until they become the major ion. However, Figure 9 shows that this behavior is dramatically changed in the region of opening and closing field lines outside the plasmasphere. The helium and hydrogen ion concentrations decline with altitude only slightly less rapidly than O^+, although the theoretical curves shown would predict otherwise. Even though the field lines are not permanently open to allow freest escape, upward movement and loss of the light ions from the ionosphere seems apparent. (We note that Banks and Holzer (1969c) have estimated the speed of efflux and influx in convecting polar field tubes.) It is important to understand that this effect does not arise from some disturbance behavior, since all the data were selected by Brinton and Grebowsky for days in which A_p was less than 22.

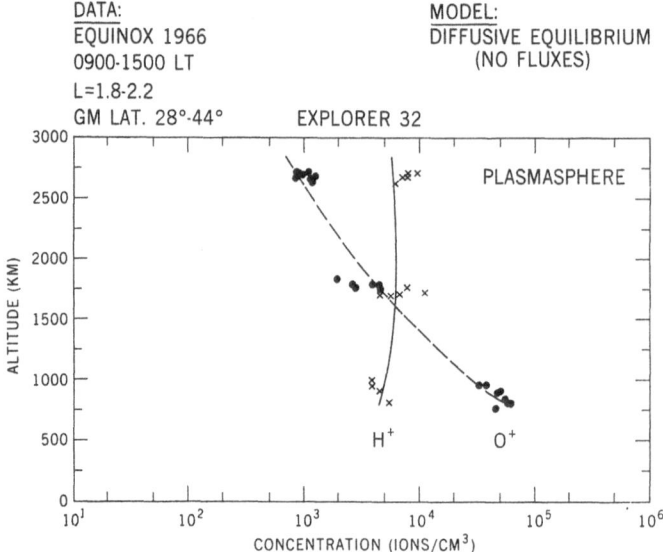

Fig. 8.   Ionic distributions within the plasmasphere from passes of Explorer 32 near equinox 1966. The curves illustrate the theoretical behavior under conditions of electrostatic diffusive equilibrium. (Brinton and Grebowsky, 1970.)

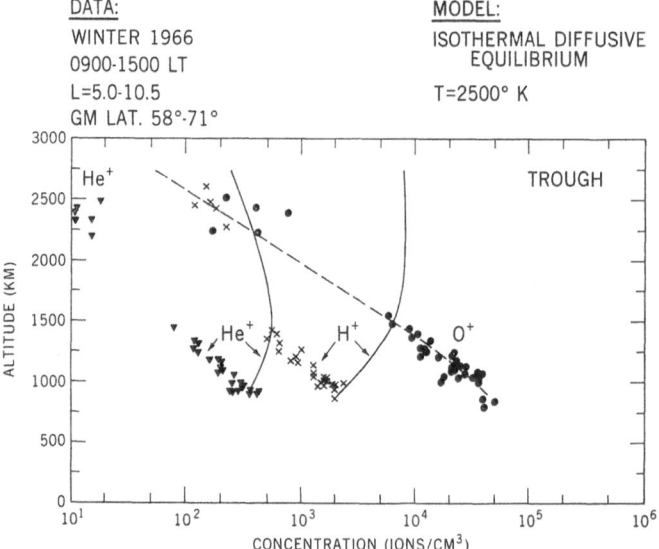

Fig. 9.   Ionic distributions within the 'trough' outside the plasmasphere from passes of Explorer 32 in winter 1966. The curves illustrate the theoretical behavior under conditions of electrostatic diffusive equilibrium. (Brinton and Grebowsky, 1970.)

7. Theory of the Polar Wind

Dungey (1961) first pointed out that open field lines would provide a channel for the escape of plasma from the earth. Bauer (1966), and more extensively, Dessler and Michel (1966), later suggested that such flow could result in the absence of H^+ at high latitudes where magnetic containment does not occur. Furthermore, in discussing the balance of terrestrial helium, Axford (1968) first pointed out that the loss of light ions from the polar ionosphere would be much enhanced over normal evaporative flow because the field generated by escaping photoelectrons would carry along light ions. (Indeed, he noted the roughly equivalent efficiency of photoionization and subsequent loss from the polar regions for He^3 and He^4. He then proposed this as a more important mechanism than evaporation, whose markedly dissimilar efficiencies for the two species had been the chief problem in understanding their equilibrium balance in the atmosphere.)

The detailed treatment of the outward polar ion flow was first begun by Banks and Holzer (1968), and further extensively developed by them in a series of three papers during the following year (Banks and Holzer, 1969a, b, c). They utilized the Euler equations of streamline flow as given in Equations (1) and (2).

$$u_j \frac{\partial u_j}{\partial r} + \frac{1}{\varrho_j} \frac{\partial P_j}{\partial r} + g(r) - \frac{eE}{m_j} = -A - \frac{q_j}{n_j} u_j \tag{1}$$

$$\frac{1}{\varrho_e} \frac{\partial P_e}{\partial r} + \frac{eE}{m_e} = 0 \tag{2}$$

n_j, m_j, ϱ_j and q_j are respectively the concentration, mass, density and production rate of the jth ion species. The first or inertial term in (1) involves the flow velocity u_j of the species. From left to right the other terms represent the driving or impeding effect of the species pressure gradient, the effect of gravity, that of the electric field which arises from the electron charge separation, and the (variable) effect of collisions with other ion and neutral species, and of the momentum to be imparted to ions newly formed at rest. The similar Equation (2) for electrons lacks terms corresponding to those in (1) where they are shown to be negligibly small. Total charge neutrality is now assumed, the perfect gas law relation between pressure and concentration is imposed (on the assumption that collisions are frequent), and electron ion and temperature models are adopted. The resulting solution of ion velocity as a function of altitude is multi-branched with one branch becoming supersonic for the condition that the plasma pressure be negligibly small at infinity. Other solutions which never become supersonic (though they rise to a maximum velocity) correspond to a definite boundary pressure of the plasma at great distance (as along a connected field line).

The resulting H^+ velocities are shown in Figure 10 (Banks and Holzer, 1969c) for characteristic daytime models with several ion temperatures. Note that the supersonic transition, or critical point, given by the '+' marks at less than 5 km/sec on the horizontal velocity scale, correspond to altitudes in the 3000 to 6000 km range on

the vertical scale. The model distributions are such that in the region below the
critical point the free path of hydrogen ions in the oxygen ion medium is always less
than the oxygen ion scale height. Thus, applicability of the hydrodynamic formulation
is assured. Figure 11, from the same paper, displays densities plotted horizontally
against altitude for a more elaborate temperature model. They reveal that the hydrogen
ion concentration has been reduced to less than atomic oxygen for all altitudes below

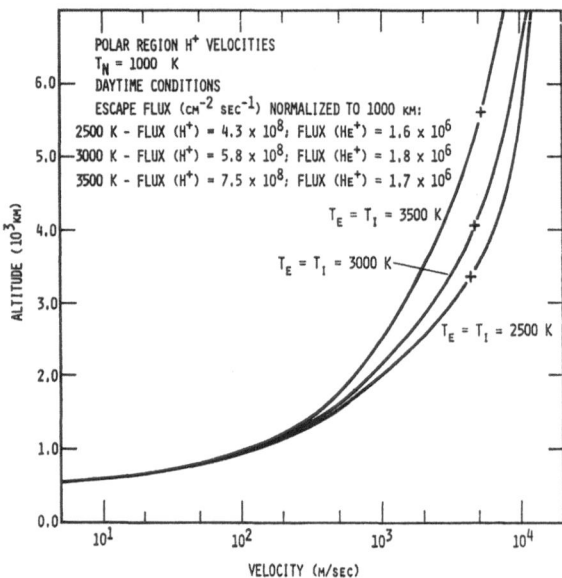

Fig. 10.   Daytime polar region H$^+$ velocity profiles from hydrodynamic theory for neutral temperature of 1000 K and several ion-electron temperatures. Units of flux are ions cm^{-2} sec^{-1}. (After Banks
and Holzer, 1969c.)

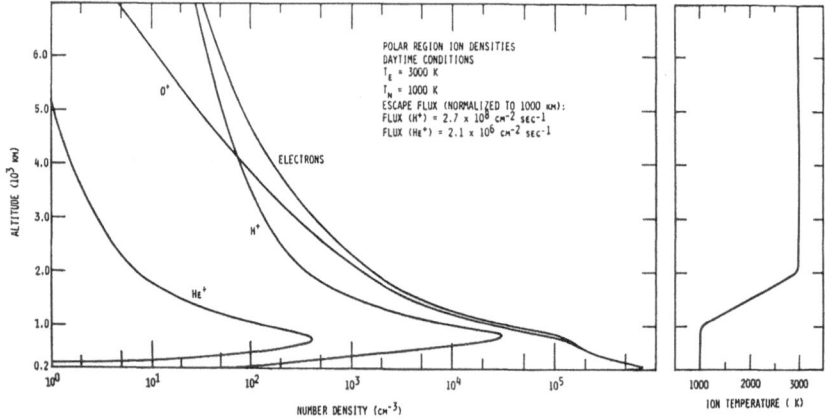

Fig. 11.   Daytime polar region ion composition profiles from hydrodynamic theory for neutral
temperature of 1000 K, electron temperatures of 3000 K and ion temperature profile as shown.
(After Banks and Holzer, 1969c.)

4000 km. These figures are sufficient to sketch the general character of the solutions obtained from integration of the hydrodynamic flow equations.

An alternative, and complementary treatment of the polar wind has recently been published by Lemaire and Scherer (1969, 1970a). Theirs is a kinetic approach in which they first develop expressions for the electric field potential in the exosphere for the Poisson equation. They impose the requirement that the electron charge density effectively cancel the positive ion charge density. Equality of positive and negative fluxes is also required everywhere in the exosphere. From the electric potential distribution and the geometry of the assumed dipole open magnetic field the total potential energy for each kind of particle is completely defined up to 20000 km. A maxwellian velocity distribution function is introduced at the exobase (2000 to 3000 km), and generalized to take account of bulk velocity flow. Liouville's theorem is then invoked to obtain the velocity distribution of the thermal particles in the collisionless exosphere. Appropriate integrations over velocity space yield the density, flux, pressure tensors, and transverse and longitudinal temperatures. Now the classical electric force arising from charge separation has often been taken as one-half the gravitational force acting on the O^+ ions in rough estimates of plasma escape effects. The results of Lemaire and Scherer show that this is drastically reduced – for example, by one-half – upon introduction of as few hydrogen ions as 10% of the total.

In Figure 12 we see Lemaire and Scherer's calculation (1970b) of the density of

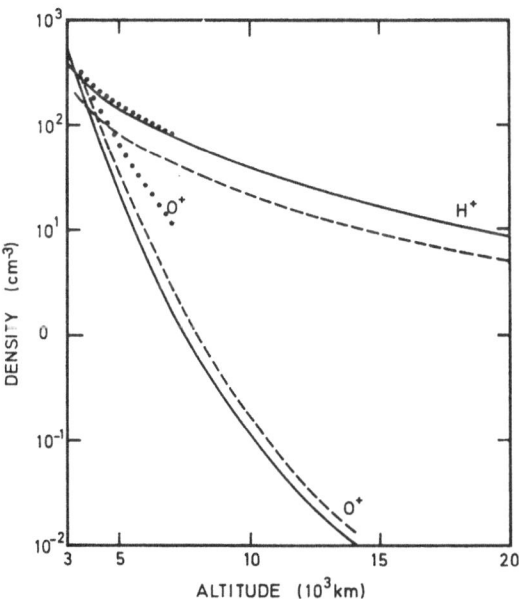

Fig. 12. Ion concentration profiles from kinetic theory for ion and electron temperature of 3000 K above a 3000 km exobase. Solid curves represent the summer daytime exosphere with photoelectron flux; dashed curves correspond to winter without such flux. Hydrodynamic results of Banks and Holzer are shown as dotted curves. This figure and Figure 13 were prepared to compare the results from kinetic theory with the models from hydrodynamic theory; they are not necessarily most representative of any actual ion-exosphere properties. (Lemaire and Scherer, 1970b.)

ions plotted as a function of altitude above a 3000 km exobase which is matched to conditions in one of the daytime hydrodynamic flow calculations of Banks and Holzer. The ion and electron temperatures are 3000 K. The solid lines illustrate the summer exosphere with a photoelectron flux of 3.2×10^8 electrons cm^{-2} sec^{-1} corresponding to the hydrodynamic calculation; the dashed lines correspond to winter conditions without such flux. The hydrodynamic results are plotted as dotted lines and are seen to be in excellent agreement for the H$^+$ distribution. However, they show a much greater scale height and density for O$^+$. In Figure 13 bulk velocities are plotted for

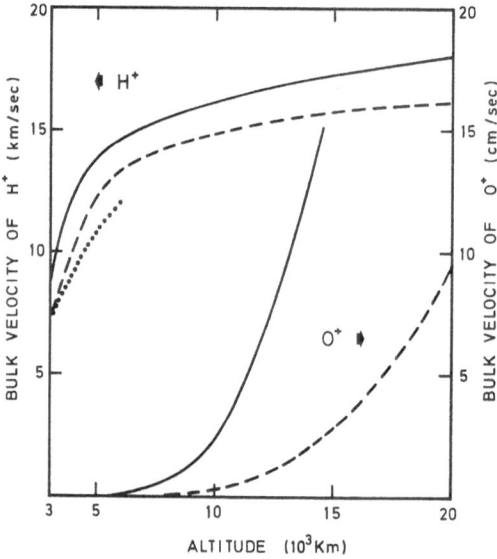

Fig. 13.   Ion bulk velocity distributions calculated from kinetic theory for the conditions of Figure 12. The dotted curve is a plot of the hydrodynamic results of Banks and Holzer.
(Lemaire and Scherer, 1970b.)

the same conditions as a function of altitude against the vertical scale on the left for hydrogen ions, and against the scale on the right for oxygen ions. The hydrogen velocities, although not identical for the two methods, are roughly comparable and certainly supersonic (whether or not there is high photoelectron flux). The oxygen velocities plotted are all very low, of order cm/sec, while those of Banks and Holzer in this particular model are too large to plot. However, it must be stressed that high oxygen flow is not essential to the hydrodynamic formulation, and does not appear in the latest models.

To summarize the theory at this time, we would say that even though the kinetic calculation neglects collisions above the exobase, whereas the hydrodynamic statement includes them, both treatments testify to the strength and supersonic velocity of the polar wind. We must be indebted to the hydrodynamic theorists for predicting these effects. We are indebted to the kinetic theorists for substantiating the polar wind concept and emphasizing the lack of appreciable oxygen flow. Further comparison of the methods must hinge on detailed equivalence of conditions in the same region, as at the exobase. Both approaches may be expected to undergo refinement as anisotropic pressure and temperature models are introduced to conform better with observation.

8. Some Observational Evidence for the Polar Wind

Of course one of the critical experiments to be done in assessing polar wind theories is the measurement of the flux velocity of hydrogen or other ions in the polar regions. Hoffman's magnetic mass spectrometer on the polar-orbiting Explorer-31 satellite has provided such measurement (Hoffman, 1970). The spacecraft cartwheeled along its orbit so that the ambient ion gas entered the instrument under cyclic ram or rarefaction condition as the spectrometer axis pointed either along, or away from, the orbital velocity vector. A typical result taken from his paper is plotted in Figure 14

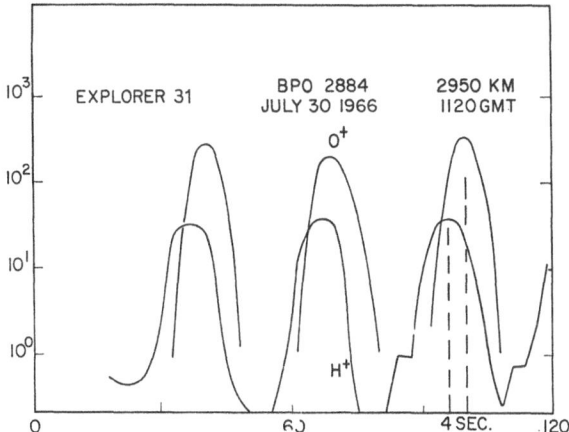

Fig. 14.   Mass spectrometer data from satellite Explorer 31 which show direct evidence of the polar wind. Ion concentration is plotted versus time showing the phase difference in the roll modulation maxima of H+ and O+. (After Hoffman, 1970.)

where more than two decades of ion current modulation are seen for both hydrogen and oxygen. As the spectrometer looks down, and then ahead, the hydrogen peak leads the oxygen by four seconds out of an approximate 30 second cycle. It is interesting to note that the phase difference, and the inferred supersonic flow, is effectively constant over the geomagnetic latitude range for the traverse from 75° to 80° North on this summer day at 2950 km altitude.

Hoffman reports typical vertical (supersonic) flow velocities for H^+ of 10 to 15 km/sec at altitudes above 2500 km. The effect is found where oxygen is the dominant ion species: both in the narrow highly-structured winter polar region between 75° and 85° North geomagnetic latitude, and also in the relatively unstructured summer regions north of 60° geomagnetic latitude. The total ion concentration in the summer regions is almost two orders of magnitude below that at lower latitudes where H^+

is the dominant species. Thus, we have a very clear demonstration of strong supersonic flow over broad geographic areas.

Ultraviolet observations from the OGO-4 spacecraft over the polar regions have been used by Meier (1970) to show the influence of the polar wind. His data, reproduced in Figure 15, depict (from top to bottom) (a) the intensity of the oxygen 1304 Å airglow lines from below, (b) that for Lyman-α from below and (c) that for Lyman-α from above, as a polar traverse is made up to 86° eccentric dipole latitude and back.

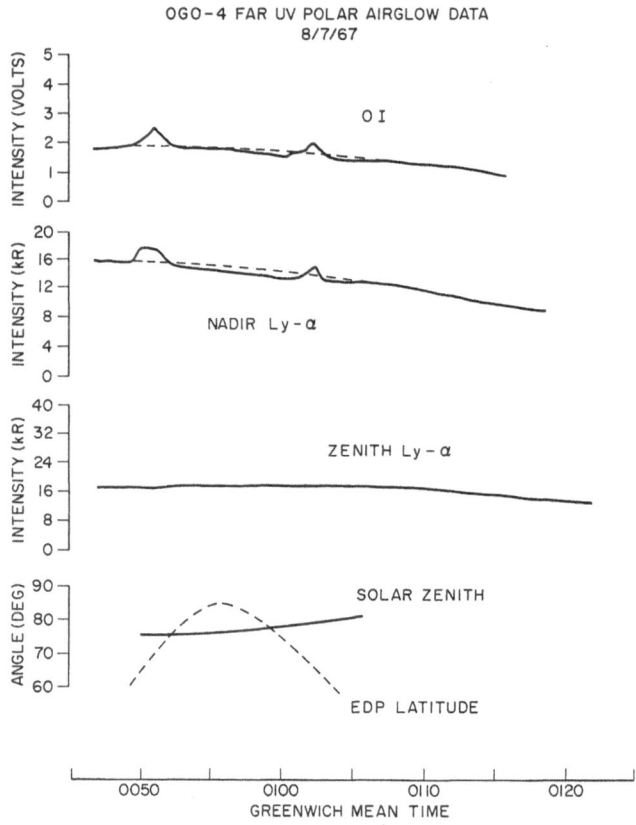

Fig. 15.   Lyman-α and O I airglow over the North Pole region. Increases in the O I and nadir Lyman-α plots near 0050 and 0102 are auroras. The dashed lines show the airglow level, if no polar depressions are present. (After Meier, 1970.)

Note that there is a broad region where the emission is depressed below the inter-polated intensity level (signified by a dashed line) for emissions from below, but that no such depression is seen in the Lyman-α from above. This depression phenomenon is taken to reflect a depletion in the usual abundance of atomic hydrogen, and was predicted by Banks and Holzer (1969c). (Other records taken from OGO-4 reveal that the depression phenomenon is *not* exhibited in the 1350–1550 Å emission in the

N_2 Lyman-Birge-Hopfield bands from below.) Meier's plot of the extent of the depressed emission along trajectories across the north polar region on August 6, 1967 is shown in Figure 16. There is a pronounced shift of the pattern off the pole toward the early morning side of the earth. Figure 17 reveals the same offset for a ten-day period. (All squares and circles – open or blackened – mark the limits of the depressions, while the crosses designate the shallow depression maximums.)

Meier has suggested that the probable cause of reduced oxygen emission at 1300 Å

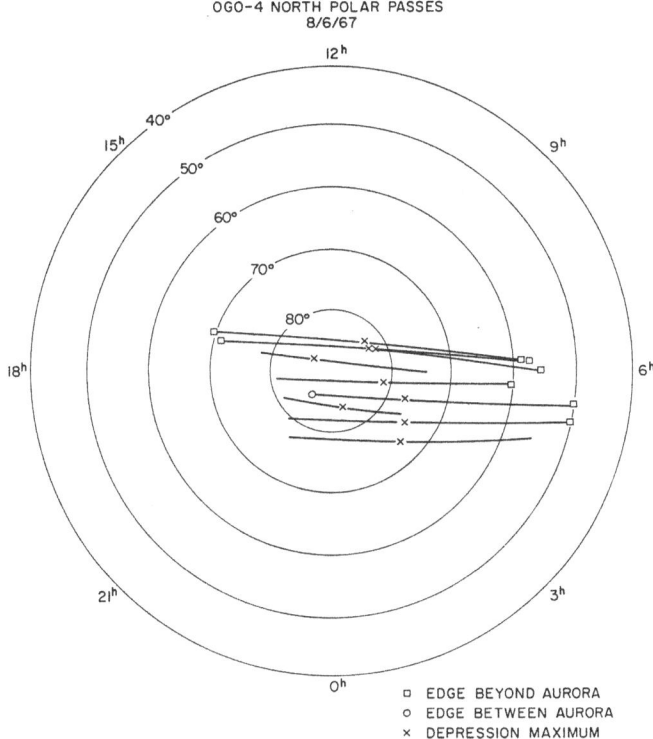

Fig. 16. Projections of OGO-4 orbits on an eccentric dipole latitude-time polar plot for depression observations. A projection with no symbol at end-point indicates that the depression boundary was obscured by an aurora. (After Meier, 1970.)

is the lack of exciting photoelectrons, since they can escape along open field lines in accordance with the calculations of Hanson (1963) and of Prasad (1969).

On the other hand, the Lyman-α depression signifies the influence of the polar wind. This is graphically displayed by Meier in Figure 18 which is divided into three altitude intervals: (I) the region of diffusion with flow, (II) the region where the process of charge-exchange with O^+ emphasized by Hanson and Ortenburger (1961) is important, and (III) the exosphere. In the regions of low plasma pressure the polar wind acts to deplete the proton concentration in interval II. The resulting loss of atomic hydrogen from charge-exchange with O^+ serves to increase the net flux from

I to II and to reduce the density in I. Thus, if the proton loss were large it could cause a distribution of atomic hydrogen like the dashed curve. If the depletion were as large as 37% Meier would predict a corresponding reduction of about 10% in the intensity of Lyman-α scattered upward. This is sometimes observed, although it is larger than the depression shown in Figure 15. This description of the depletion of hydrogen from the thermosphere by the polar wind requires that the polar flux, estimated by Banks and Holzer (1969c) at 2–7×10^8 cm^{-2} sec^{-1} for polar summer, be large enough

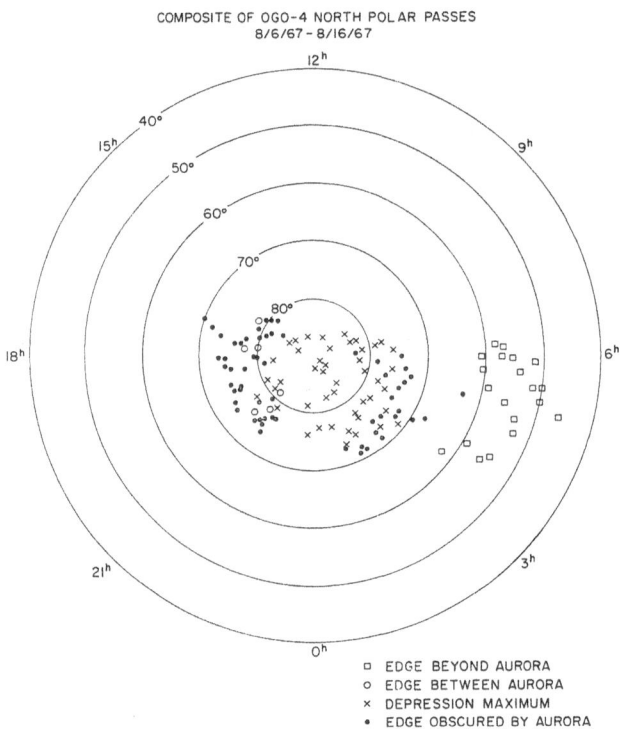

Fig. 17.   Composite of OGO-4 North polar passes for the period 8/6/67–8/16/67 showing extent of intensity depressions. (After Meier, 1970.)

to overcome any compensating lateral flow, which has been estimated to be of order 10^8 cm^{-2} sec^{-1}. The detailed calculation of the flow exchange has not been made.

In conclusion, it seems apparent that polar wind theory has moved ahead of the available data. What are most needed are broad morphological studies, especially of the supersonic flow itself, for correlation both with the theory and with the other indirect observations in the polar regions.

Acknowledgement

Many useful discussions with R. R. Meier are gratefully acknowledged.

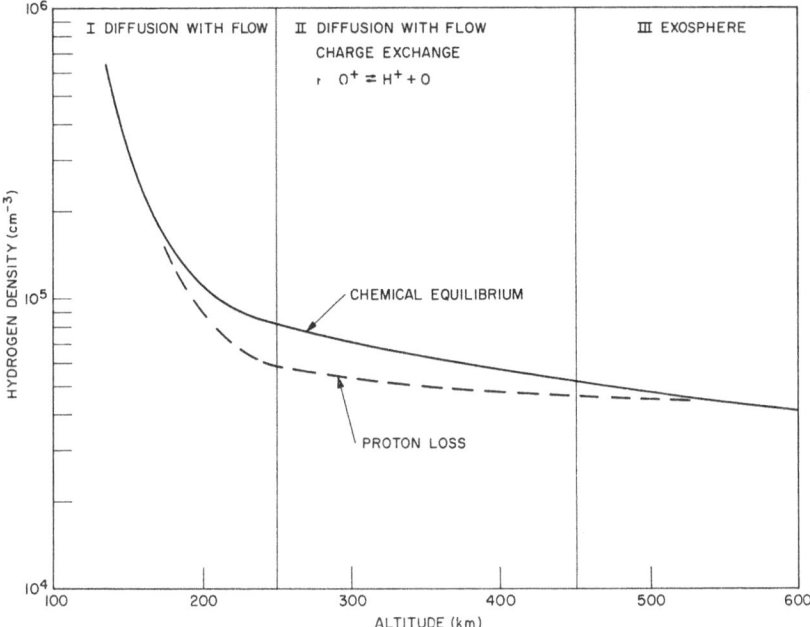

Fig. 18. Hydrogen density as a function of altitude for the normal nonpolar regions (solid line) and estimated polar distribution (dashed line) resulting from proton loss via polar wind. (After Meier, 1970.)

References

Axford, W. I.: 1968, *J. Geophys. Res.* **73**, 6855.

Axford, W. I. and Hines, C. O.: 1961, *Can. J. Phys.* **39**, 1433.

Banks, P. M. and Holzer, T. E.: 1968, *J. Geophys. Res.* **73**, 6846.

Banks, P. M. and Holzer, T. E.: 1969a, *J. Geophys. Res.* **74**, 3734.

Banks, P. M. and Holzer, T. E.: 1969b, *J. Geophys. Res.* **74**, 6304.

Banks, P. M. and Holzer, T. E.: 1969c, *J. Geophys. Res.* **74**, 6317.

Bates, D. R. and Nicolet, M.: 1950, *J. Geophys. Res.* **55**, 301.

Bates, D. R. and Patterson, T. N. L.: 1961a, *Planetary Space Sci.* **5**, 257; 1961b, *Planetary Space Sci.* **5**, 328.

Bauer, S. J.: 1966, *Electron Density Profiles in Ionosphere and Exosphere* (ed. by J. Frihagen), North-Holland Publishing Co., p. 387.

Bjuro, E. D., Zhukov, A. P., Martynkevitch, G. M., and Shvidkovsky, E. G.: 1969, *Space Res.* **9**, 501.

Brinkman, R. T.: 1970, *Planetary Space Sci.* **18**, 449.

Brinton, H. C. and Grebowsky, J. M.: 1970, private communication: also 1970, *Trans. Amer. Geophys. Union* **51**, 397 (Abstract).

Chamberlain, J. W.: 1963, *Planetary Space Sci.* **11**, 901.

Chamberlain, J. W. and Campbell, F. J.: 1967, *Astrophys. J.* **149**, 687.

Dessler, A. J. and Michel, F. C.: 1966, *J. Geophys. Res.* **71**, 1421.

Donahue, T. M.: 1967, *International Dictionary of Geophysics*, Pergamon Press, p. 733.

Donahue, T. M. and Thomas, G. E.: 1963, *J. Geophys. Res.* **68**, 2661.

Dungey, J. W.: 1961, *Phys. Rev. Letters*, **6**, 47.

Hanson, W. B.: 1963, *Space Res.* **3**, p. 282.

Hanson, W. B. and Ortenburger, I.: 1961, *J. Geophys. Res.* **66**, 1961.

Hanson, W. B. and Patterson, T. N. L.: 1963, *Planetary Space Sci.* **11**, 1035.

Hodges, R. R. Jr. and Johnson, F. S.: 1968, *J. Geophys. Res.* **73**, 7307.

Hoffman, J. H.: 1970, *Int. J. Mass Spectrom. Ion Phys.*, to be published.

Johnson, F. S. and Fish, R. A.: 1960, *Astrophys. J.* **131**, 502.

Keating, G. M. and Prior, E. J.: 1968, *Space Res.* **8**, 982.

Kockarts, G. and Nicolet, M.: 1963, *Ann. Geophys.* **19**, 370.

Lemaire, J. and Scherer, M.: 1969, *Compt. Rend. Acad. Sci. Paris* **269**, 666.

Lemaire, J. and Scherer, M.: 1970a, *Planetary Space Sci.* **18**, 103.

Lemaire, J. and Scherer, M.: 1970b, private communication.

Mange, P. and Meier, R. R.: 1970, *J. Geophys. Res.* **75**, 1837.

Mange, P., Purcell, J. D., and Tousey, R.: 1960, *Astron. J.* **65**, 54 (Abstract).

McAfee, J. R.: 1967, *Planetary Space Sci.* **15**, 599.

Meier, R. R.: 1969, *J. Geophys. Res.* **74**, 3561.

Meier, R. R.: 1970, *J. Geophys. Res.* **75**, 6218.

Meier, R. R. and Mange, P.: 1970, *Planetary Space Sci.* **18**, 803.

Meier, R. R., Weiss, D. M., and Mange, P.: 1970, *J. Geophys. Res.* **75**, 4224.

Nishida, A.: 1966, *J. Geophys. Res.* **71**, 5669.

Patterson, T. N. L.: 1966, *Planetary Space Sci.* **14**, 425.

Prasad, S. S.: 1969, *J. Geophys. Res.* **74**, 4772.

Reber, C. A., Cooley, J. E., and Harpold, D. N.: 1968, *Space Res.* **8**, 993.

Thomas, G. E.: 1963, *J. Geophys. Res.* **68**, 2639.

Thomas, G. E., Barth, C. A., and Hord, C. W.: 1970, *Trans. Amer. Geophys. Union* **51**, 587 (Abstract).

Tinsley, B. A. and Meier, R. R.: 1972, International Symposium on Solar-Terrestrial Physics, Leningrad.

Weller, C. A., Young, J. M., Johnson, C. Y., and Holmes, J. C.: 1970, private communication; also 1970, *Trans. Amer. Geophys. Union* **51**, 367 (Abstract).

ELECTRIC FIELDS AND THEIR EFFECTS IN THE IONOSPHERE

(Summary of Observations)

G. HAERENDEL

Max-Planck-Institut für Physik und Astrophysik
Institut für extraterrestrische Physik, Garching bei München

Abstract. The quasi-static electric field is a fundamental parameter of the physical processes in the magnetosphere and ionosphere. Yet it is difficult to measure. Only five years ago we had to rely netirely on indirect evidence, mostly derived through relating the magnetic perturbations on the ground to an overhead ionospheric current (Maeda, 1955; Axford and Hines, 1961). During the last few years a number of different techniques have been developed and successfully applied which provide a quickly. growing body of data on the electric field or the plasma convection (Völk and Haerendel, 1970) All these techniques are of limited applicability. Simultaneous observations using two or more different methods are highly desirable at the present stage for experimental cross checks and for obtaining complementary information.

This review is devoted to a brief survey of the observational methods and the most significant results. A few conclusions from these data on the electromagnetic coupling of the ionosphere and magnetosphere and on the magnetospheric convection will be added.

1. Methods

1.1. SURVEY OF TECHNIQUES

The methods for measuring the electric field, E, can be divided in two groups. Group 1 consists of only one technique, namely the direct measurement of potential differences between two Langmuir probes. Group 2 derives the electric field from measurements of the plasma convection velocity, v. In the absence of neutral air drag and strong pressure differences E and v are related by:

$$E + \frac{1}{c} v \times B \cong 0. \tag{1}$$

Whereas Group 1 can in principle be used to measure also the parallel component of E, we get only indirect information on $E_{\parallel}$ from the data of Group 2. However, by making assumptions on the interaction of charged particles with each other, with neutral particles or plasma waves the motion along magnetic field lines can be used to derive $E_{\parallel}$.

Table I presents a survey on the different techniques. Only a few comments shall be added.

The double-probe technique depends very much on the realization of symmetric contact potentials between probe and plasma. Asymmetries in geometry, work function and in the photo-emissivity have to be carefully avoided (Aggson and Heppner, 1964, 1965; Fahleson, 1967; Fahleson *et al.*, 1968; Mozer and Bruston, 1967). For probes flown in or below the ionosphere this has been achieved. Asymmetric instruments as that of Kavadas and Johnson (1964) are not easy to interpret.

It is doubtful whether the double-probe technique is equally applicable in the dilute

TABLE I

Techniques for measuring d.c. electric fields

1. *Double probes*

2. *Measurements of plasma drifts*
   2.1.   *Electron drifts*
   2.1.1. Radar echoes
   2.1.2. Quadripole probe
   2.1.3. Whistler duct tracing
   2.1.4. Electron beam

   2.2.   *Ion drifts*
   2.2.1. Retarding potential analyzer, split Langmuir probes
   2.2.2. Electrostatic flux meters
   2.2.3. Barium plasma clouds
   2.2.4. Asymmetry of proton velocity distribution

magnetospheric plasma at great altitudes outside the plasmasphere where the photo-electron sheath surrounding the satellite can reach a thickness of several meters. Even with booms of a length of several tens of meters deviations from perfect symmetry must not exceed 10^{-3} for a successful measurement. The question of probe potential is of critical importance for all the measurements in the magnetosphere (Whipple, 1970; Grard and Tunaley, 1970).

All probes flown with space vehicles have the disadvantage that the field induced by the motion of the vehicle, $(1/c) \mathbf{v}_s \times \mathbf{B}$, must be subtracted from the measured total field. At medium and low latitudes the quantity to be measured is typically one order of magnitude less. This disadvantage is avoided by flying the double probe on balloons underneath the ionosphere. This method was suggested by Kellogg and Weed (1969) who demonstrated that the horizontal electric field at heights of 30–40 km should be closely related to the ionospheric field of scales on the order of 100 km or more (Atkinson *et al.*, 1969; Mozer and Serlin, 1969). However, this method suffers from perturbations by local horizontal electric fields which cannot always be separated unambiguously.

Radar echoes incoherently or coherently scattered at ionospheric irregularities provide measurements of electron velocities by the Doppler effect (2.1.1.).

Woodman and Hagfors (1969) suggested a method of monitoring vertical drifts in the F layer by measuring the autocorrelation of the backscattered signals of double radar pulses. If the beam is transverse to the magnetic field an accuracy of a few meters per second can be achieved. Also the measurements of the Doppler shift of single incoherently scattered radar pulses as obtained with the facilities in Jicamarca, Arecibo, Nançay and Millstone Hill bear a potential wealth of information on the electric field.

Balsley (1969) studied the coherently scattered radar echoes from certain electron-density irregularities in the equatorial electrojet and derived horizontal electron drift velocities. Several other types of ionospheric irregularities can be detected by radio techniques and apparent velocities be derived. In some cases, however, wave velocities are measured clearly, in other cases the physical process has still to be explored.

The method of studying the mutual impedance of two electric dipoles (one active, one passive) for deriving plasma drift velocities (2.1.2) was suggested by Storey *et al.* (1968). The effect is an additional damping of the resonances at the lower hybrid resonance frequency. Rocket data showed the expected effect; however the quantitative applicability of the method has still to be demonstrated.

Nose whistlers can be analyzed for the electron density at lower latitudes up to altitudes of several earth radii (Carpenter, 1963, 1966; Angerami and Carpenter, 1966). In some cases changes of the nose frequency can be attributed to the radial displacements of the magnetic flux tubes providing the whistler duct (Carpenter and Stone, 1967) (2.1.3). This is one of the few techniques that yielded direct data on magnetospheric plasma drifts. Motions in longitude are much more difficult to derive with this technique.

The measurement of the ion current of the thermal plasma above the E layer of the ionosphere provides a simple means of measuring the plasma drift. This can be achieved by retarding potential analyzers (Freeman, 1968; Pfister, 1969; Spenner, 1970) or two parallel plane Langmuir probes (Bering *et al.*, 1970) (2.2.1), which show normally spin-modulated ion current signals. In the magnetosphere the technique is affected very seriously by the relatively high positive probe potential in sunlight or even higher negative potential in the earth's shadow.

Historically the first attempt to measure the electric field in space was made by Gdalevich (1964) and Imyanitov *et al.* (1964) by using the so-called field mill or electrostatic flux meter (2.2.2). Knott (1970) made use of the rocket spin to measure the electrostatic field on the surface of the rocket. These measurements, however, give only indirect evidence of the ambient field. The plasma drift leads to distortions of the plasma sheath surrounding the space vehicle and thus to a strongly asymmetric field. Therefore, the direction of the plasma wake can be safely deduced. The plasma density and temperatures and some theory, however, are needed to derive the magnitude of the ion drift velocity from the data (Gurevich, 1964; Alpert *et al.*, 1965; Taylor, 1967; Liu, 1969).

The barium plasma cloud technique of measuring ionospheric plasma drifts is very simple and relatively reliable (Haerendel *et al.*, 1967; Hacrendel and Lüst, 1968b) (2.2.3). Its disadvantage is the very limited applicability being restricted to twilight conditions. Lasers may be used to extend the observation time. In the higher magnetosphere the tracing of the natural plasma drift is not as easily achieved as in the ionosphere because of the long time scales involved in the momentum transfer from the natural to the barium plasma (Scholer, 1970; Haerendel and Lüst, 1970).

In view of the difficulties opposing a reliable measurement of the electric field in the outer magnetosphere the study of the drift of energetic electrons or ions offers a promising alternative. Regarding the low ratio of drift velocity to total velocity of the electrons, natural electrons do not offer sufficient anisotropy to detect the drift velocity. However, a narrow beam of electrons emitted by a gun on board a satellite can be distinguished from the natural background and can be detected upon return to the satellite after one gyroperiod. The drift velocity of the ambient plasma shows up in a spatial displacement of the returning beam (2.1.4). The feasibility of such an active measurement was demonstrated by Melzner and Völk (1970) in a proposal for the ESRO geostationary satellite. At low altitudes the gyroradii of electrons are too small. A similar technique using ions could be tried.

The natural protons near 1 keV in the outer magnetosphere offer another possibility to derive the drift from the asymmetry of their distribution over the phase angle at constant pitch-angle (2.2.4). If the latter is chosen to be 90°, the anisotropy, A_E, due to

$$\mathbf{v}_E = c \, \frac{\mathbf{E} \times \mathbf{B}}{B^2} \tag{2}$$

is related to the gradient in the energy dependence of the differential unidirectional flux, j, and the total velocity, v, by

$$A_E = \left(1 - \frac{\partial \ln j}{\partial \ln E}\right) \frac{4v_E}{v}. \tag{3}$$

A similar anisotropy, A_G, is introduced by spatial gradients of j:

$$A_G = 2R_g \, \frac{|\nabla j|}{j}. \tag{4}$$

However, separation of both causes of anisotropy can be achieved by comparing A_E and A_G at different energies, since $A_G/A_E \sim E$. (R_g is the gyroradius.)

Certain indirect conclusions on the large scale quasi-static electric field in the magnetosphere and tail can be drawn from the interpretation of some significant intensity and spectral variations. This will be discussed in Section 2.4.

This survey of techniques shall be concluded with a few remarks on attempts to derive the *parallel electric field*. Direct measurements of Mozer and Bruston (1967) with double probes yielded data suggesting unexpectedly high electric field components parallel to the magnetic field of 20 mV/m. Mostly, however, the parallel field appears to be smaller by orders of magnitude. The presence of a small, but detectable $E_\parallel$ is sometimes indicated by vertical motions of barium clouds. However, the relation to the electric field is controversial, since assumptions on the parallel mobility of the charged particles must be made in order to derive $E_\parallel$. Plasma instabilities creating turbulence and thus reducing the effective conductivity develop in the presence of strong currents in a low density plasma (Swift, 1965; Coroniti, 1969; Dupree, 1970). When the conductivity in the ionosphere is based entirely on collisions, all vertical motions observed so far do not give field strengths exceeding 10 μV/m considerably (see also Mende, 1968).

A way of probing potential differences along magnetic field lines may be provided by the measurement of minor constituents of charged particles, if their interaction properties are much different from that of the dominant plasma population. A shaped charge technique yielding fast barium atoms with velocities up to 10 km/sec and more is in development as a source of fast ionic test particles which can probe such potential differences even between conjugate points.

In the interpretation of so-called mono-energetic particle spectra and peculiar angular distributions in the auroral zone the existence of electric potential differences of several kilovolts along magnetic lines of force even at low altitudes has been frequently postulated (Chamberlain, 1969). Unambiguous measurements of these phenomena are of highest interest.

1.2. SIMULTANEOUS MEASUREMENTS WITH DIFFERENT TECHNIQUES

The relatively young age of the art to measure electric fields in space suggests very strongly to cross-check the various methods wherever possible. Few attempts have been made. Wescott *et al.* (1969) compared data from barium cloud releases with data from three axis double probes which were flown on a different rocket 10 min later in one case, 1 hr later in the other. The results were found to be consistent with each other. However, because of the strong spatial and temporal variability of the electric field, data from different flights are not conclusive.

Two Skylark rockets instrumented with a double probe, an electrostatic flux meter

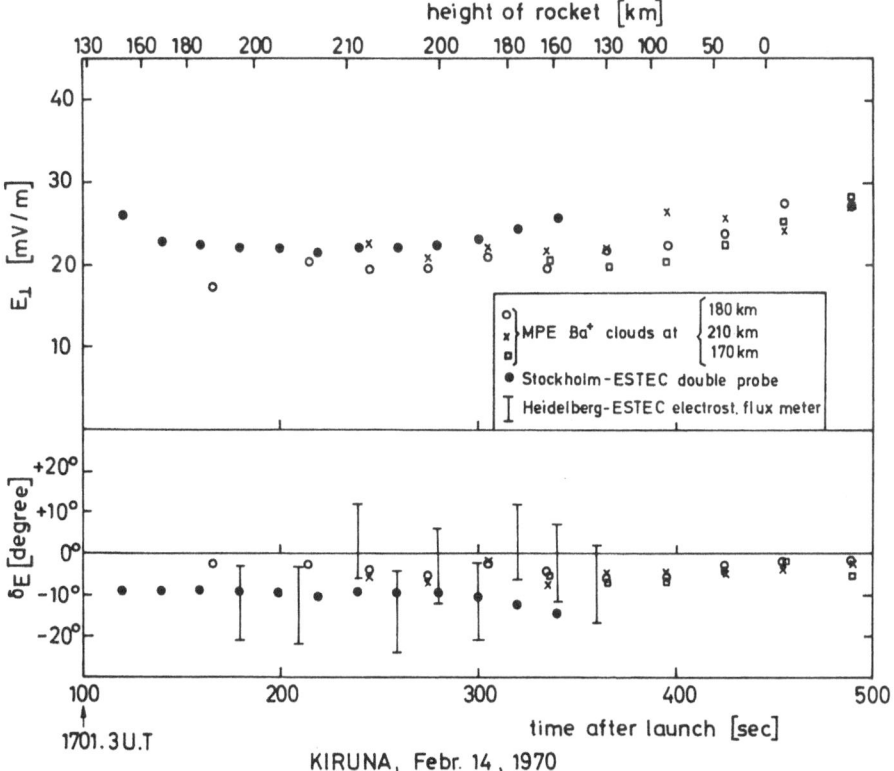

Fig. 1.   Magnitude and azimuth, δ_E, of the transverse electric field measured on a rocket flight from Kiruna simultaneously with an electrostatic double probe, barium ion clouds and an electrostatic fluxmeter (Fahleson *et al.*, 1970). δ_E is measured in a plane transverse to **B** and counted positive from north to east. A positive horizontal magnetic perturbation of 45γ was observed on the ground during the rocket flight.

and barium containers were launched from ESRange in early 1970 (Fahleson *et al.*, 1970). Data from the first flight are compared in Figure 1. The barium cloud and double probe data are in very good agreement (within 10%) on the upper parts of the trajectory where the least perturbations are expected.

Another comparison of two techniques was carried out by Bering *et al.* (1970). Electric fields derived from the ion current as measured with a split Langmuir probe agree very well with direct double probe measurements on the same rocket.

2. Results

The data on electric fields have grown so much that their full appreciation in a short review is not possible. In the following sections we shall only attempt to give a survey of the type of data that exist omitting many details and without drawing many conclusions about their meaning.

2.1. DATA FROM LOW AND MEDIUM LATITUDES

Data on electric fields in the ionosphere at low and medium latitudes were obtained by the use of techniques 2.1.1. and 2.2.3. of Table I.

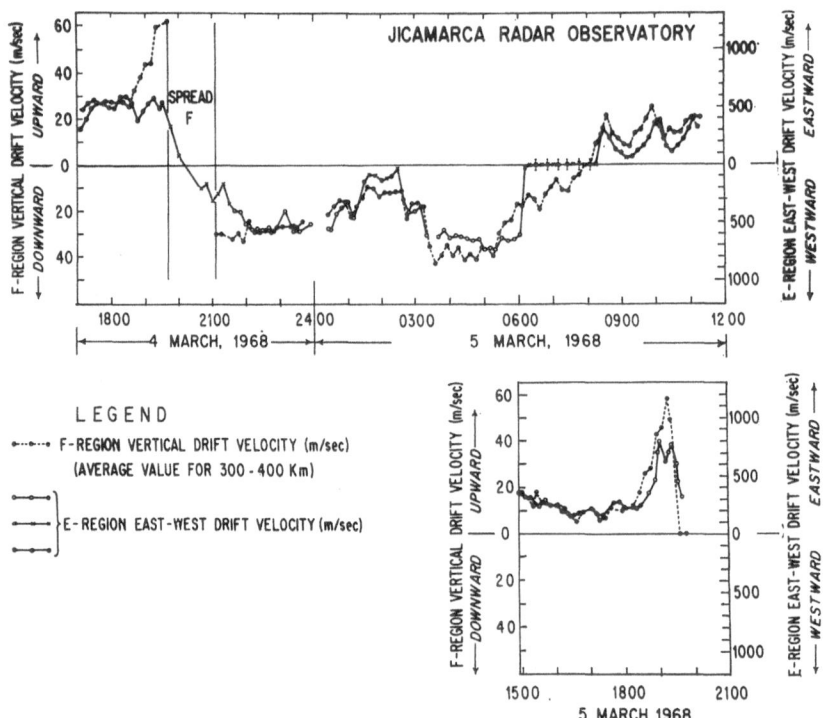

Fig. 2. Full day coverage of vertical drift velocities in the F layer and of horizontal drifts in the E layer as obtained with the radar facilities in Jicamarca (Peru) near the magnetic equator (Woodman and Balsley, 1969). Left hand ordinate refers to the F region and right hand ordinate to the E-layer drifts.

Figure 2 contains a comparison of vertical drifts in the *equatorial* F layer between 300 and 400 km as measured with the incoherent backscatter facility in Jicamarca (Peru) with horizontal drifts in the E layer obtained from radar echoes at a certain type of irregularities (Woodman and Balsley, 1969). The different scales for the vertical and horizontal drifts should be noted. The vertical electric field that would cause the horizontal drift in the region of the equatorial electrojet appears to be closely proportional to the horizontal field in the F layer with a constant of proportionality of about 20. This agrees fairly well with the interpretation of the equatorial electrojet. The data show also a rising of the F layer during the hour preceding the evening reversal.

Barium cloud releases performed in Thumba (India) by the Institut für extraterrestrische Physik and the Physical Research Laboratory in Ahmedabad confirmed the upward directed motions in the F layer before sunset of similar magnitude (Rieger, 1969; Haerendel *et al.*, 1970).

The horizontal westerly drifts exceed the vertical components by about a factor of three (Figure 3). During morning twilight we found still nighttime conditions to hold with easterly and downward directed drifts.

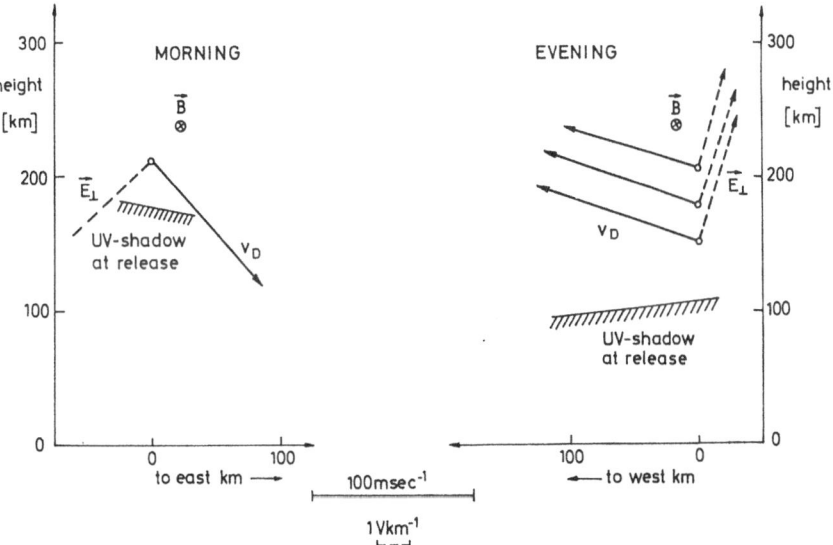

Fig. 3.   Ionospheric drift velocities and electric fields in the magnetic equatorial plane near Thumba (India) during morning and evening twilights as derived from barium plasma cloud experiments (Haerendel *et al.*, 1970).

At *mid-latitudes* all existing low altitude data were derived from barium cloud experiments (Haerendel and Lüst, 1968a). Figure 4 gives a survey of electric fields derived from barium cloud experiments of the Institut für extraterrestrische Physik (Rieger, 1970). The electric field vector has to be understood as plotted in a plane transverse to the local magnetic field vector. During twilight the electric field is typically a few mV/m. In the morning, fields in the northern hemisphere are dominated by a northward directed component, whereas southward directed fields prevail in the

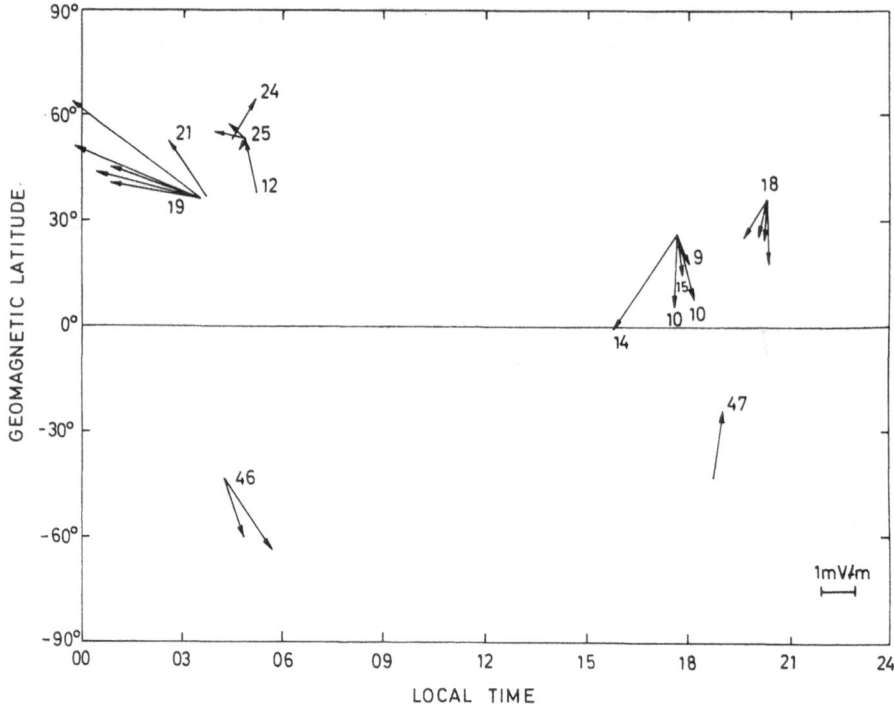

Fig. 4.   Transverse electric fields in the mid-latitude ionosphere as derived from barium plasma cloud experiments at twilight (Haerendel and Lüst, 1968a; Rieger, 1970). The vectors must be understood as plotted in a plane transverse to the local magnetic field.

evening. There are a few exceptions, typically when the field is very weak (<1 mV/m) or during strong magnetic activity (section below).

The observation that the average field direction opposes the average direction of the Sq-current was interpreted by Haerendel and Lüst (1968a) as an indication that the current is driven by the atmospheric dynamo field in the E layer. The data are in good agreement with the theory of Maeda (1955), but not with the recent Sq-model of Matsushita (1969).

It is interesting that in AFCRL barium cloud experiments by Rosenberg and Best (unpublished data) during strong magnetic perturbations ($K_p=5+$ and $7+$) the magnitude of the electric field was found not to increase.

2.2. DATA FROM HIGH MAGNETIC LATITUDES

In the light of recent theories of the magnetospheric convection pattern it is of interest to know whether the *plasmapause* also appears as a boundary with respect to the plasma motion. Gurnett (1970) reports an observation obtained with a double probe on the magnetically stabilized satellite Injun 5 where at the plasmapause the electric field changes from nearly purely corotational inside to a 15 mV/m convection field outside.

A barium cloud experiment from Kiruna (Föppl *et al.*, 1968) during a relatively quiet period offered another possibility of a comparison. Just south of the field lines defining the plasmapause as determined from OGO 2 whistler measurements (Carpenter, private communication) the ionospheric plasma drifts were directed westward. Negative bay currents and auroral arcs only 100 km to the north of the ion cloud are indicative of strong eastward directed drifts north of the plasmapause at the same time (Figure 5). The electric field inside the plasmapause was about 4 mV/m. It should be remarked that these two examples of sudden changes of the electric field at the plasmapause may not be typical for all local times.

A distinct *boundary* separating the low and medium latitude region of low electric

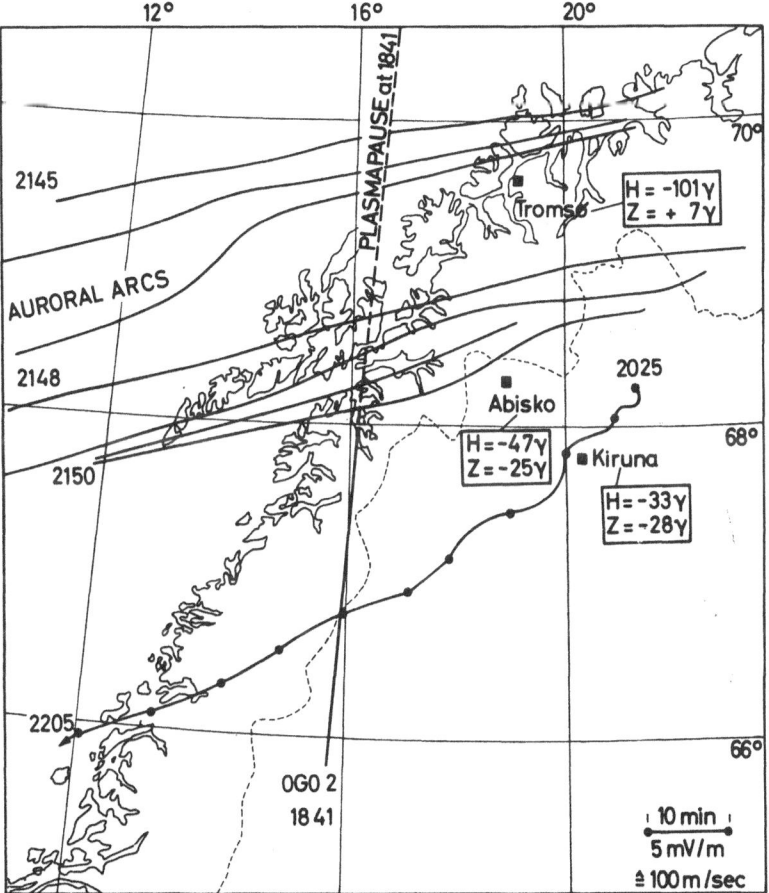

Fig. 5. Path of a barium plasma cloud immediately south of the plasmapause (Föppl *et al.*, 1968). Dots are separated by 10 minute intervals. From whistler measurements on OGO 2 (Carpenter) the plasmapause can be located on the dashed part of the OGO 2 orbit projected to the ion cloud level of 235 km. During a subsequent crossing of OGO 2 slightly west of Ireland close to the time of the ion cloud experiment, the plasmapause had moved southward in invariant latitude by about 2°. At 2030 UT a negative bay started and about an hour later some auroral arcs appeared north of Kiruna. The magnetic perturbations at 2130 UT are indicated for three stations. All times given in UT.

fields from much higher and quite irregular fields was found by Heppner (1969). Figure 6 shows this boundary as a function of local magnetic time as derived from ELF electric field channels of a double probe on the polar orbiting satellite O 5 1–10. The sudden onset of noise in the 3–30 Hz band which defines the boundary was attributed to the satellite crossing local irregularities of the electric field with typical wavelengths of less than 1 km. These irregularities may be of the same nature as the striations of barium clouds. A correlation of such irregularities with strong d.c. electric fields as postulated by Heppner (1969) was experimentally verified with another double probe on OGO 6 (Figure 7) by Maynard and Heppner (1970).

   The southern border of the region of high electric fields is shifted further to the south with increasing K_p. It resembles the lower boundary of the auroral oval, although both boundaries may not strictly coincide. This picture was confirmed by the measurements with Injun 5 (Gurnett, 1970; Cauffman and Gurnett, 1970).

   Much material was collected about the electric field in the *auroral zone*. Double probe measurements with rockets (Mozer and Bruston, 1967; Bering *et al.*, 1970; Aggson, 1969; Potter and Cahill, 1969; Fahleson *et al.*, 1970), with satellites (Gur-

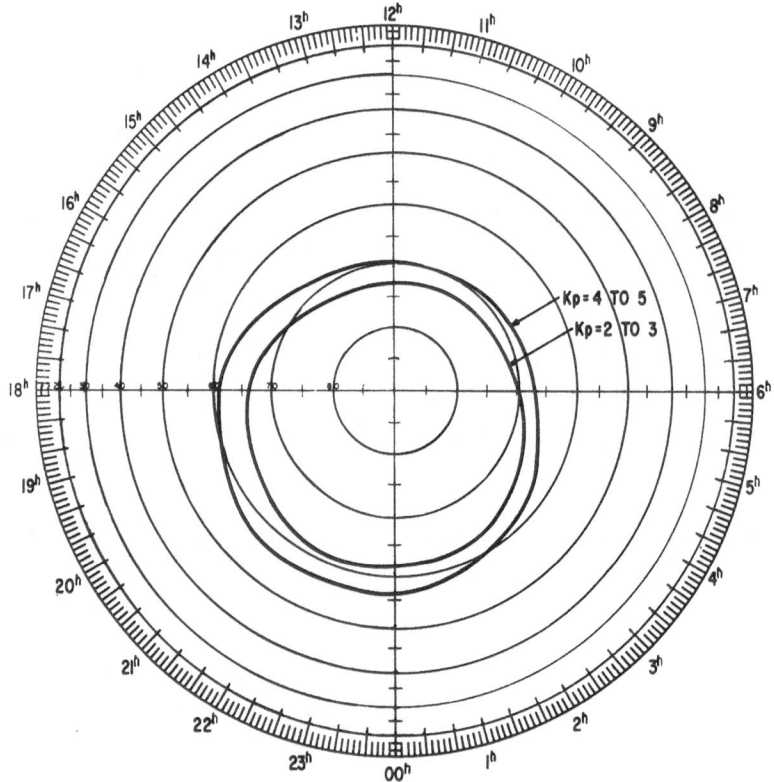

Fig. 6.   Average minimum latitude of the region of strong and irregular electric fields as derived from the 3–30 Hz electric field channel of a double probe on O 5 1–10 (Heppner, 1969). Plot in invariant magnetic latitude and magnetic local time for two intervals of K_p.

VARIATIONS IN ELECTRIC FIELDS FROM POLAR SATELLITES

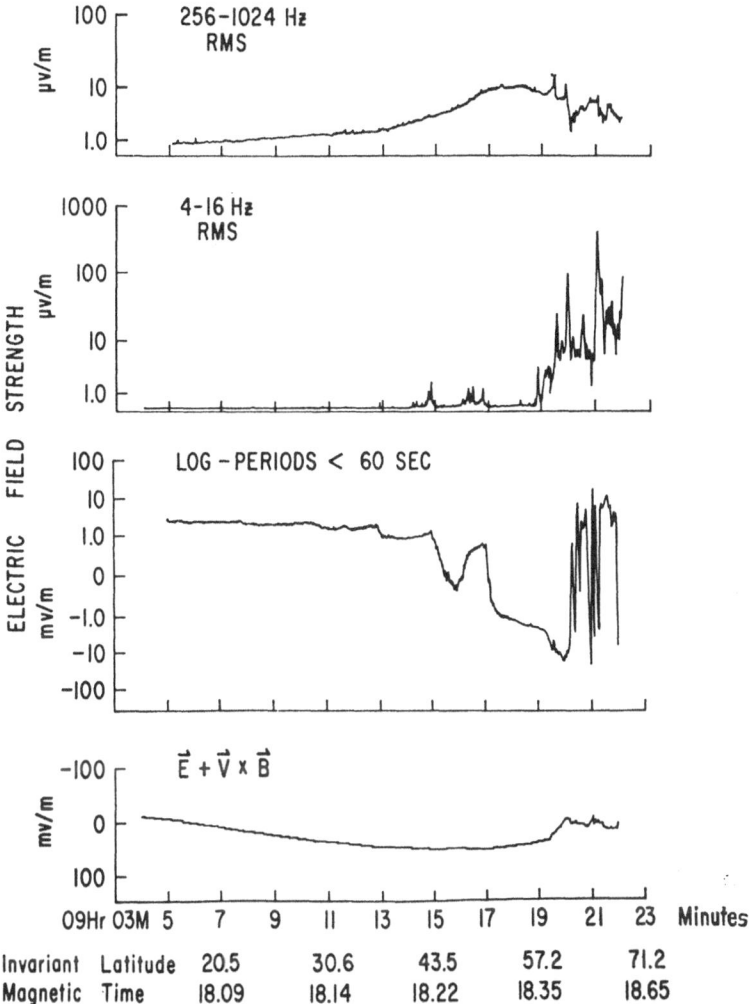

Fig. 7. Typical output of four electric field channels of a double probe on OGO 6 (Maynard and Heppner, 1970). Coincidence of regions of strong ELF fields in the 4–16 Hz band with regions of strong d.c. electric fields (lowest diagram) is demonstrated. The region of strong VLF emissions (uppermost diagram) is typically located at lower latitudes.

nett, 1970; Cauffman and Gurnett, 1970; Maynard and Heppner, 1970) and balloons (Mozer and Serlin, 1969; Mozer and Manka, 1970) as well as barium cloud data (Föppl *et al.*, 1968; Haerendel and Lüst, 1968a, 1970; Haerendel *et al.*, 1969; Wescott *et al.*, 1969, 1970) provided the major contribution. Field strengths of a few tens of mV/m are the most typical. However, much lower and also higher values up to 300 mV/m are found at times. A strong variability, spatial as well as temporal, is the most striking feature. Over scales of a few km transverse to **B**, mostly in north-south direction, or during time intervals of a few minutes, the field may change its orientation completely. The field is predominantly pointing either north- or southward. A close correlation of the former direction with positive and of the latter with negative magnetic perturbations was established (Section 3.2). Correspondingly westward directed drifts prevail during evening hours until 2300 MLT and later on drifts towards east.

A comprehensive presentation of ion cloud drifts found by the groups of the Institut für extraterrestrische Physik and of Goddard Space Flight Center (numbers preceded

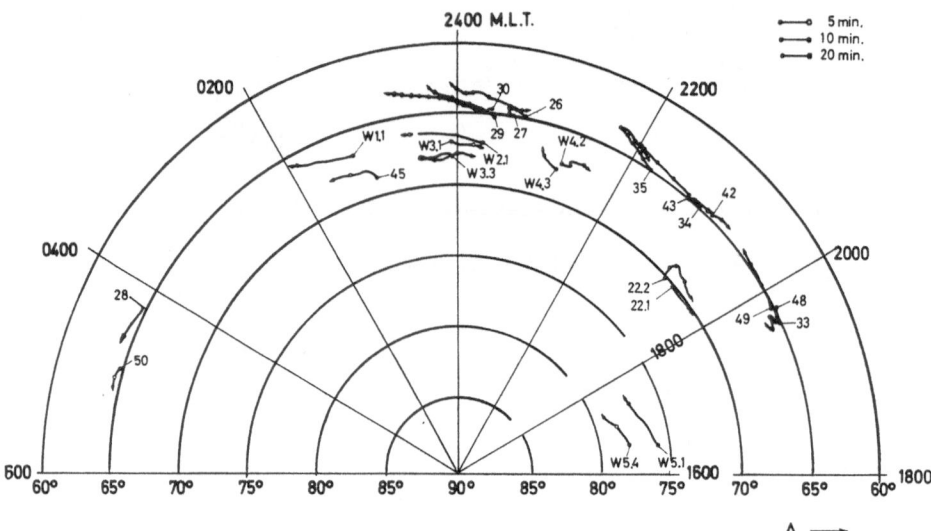

Fig. 8. Drift paths of barium plasma clouds in an invariant latitude versus magnetic local time plot (Haerendel and Lüst, 1970). Numbers indicate the different experiments. Those preceded by W were taken from Wescott *et al.* (1969, 1970). Consecutive points are separated mostly by 10 min (s. legend). The sector from 1600 to 1800 MLT is rotated by 2 hr for $\Lambda \geqslant 75°$.

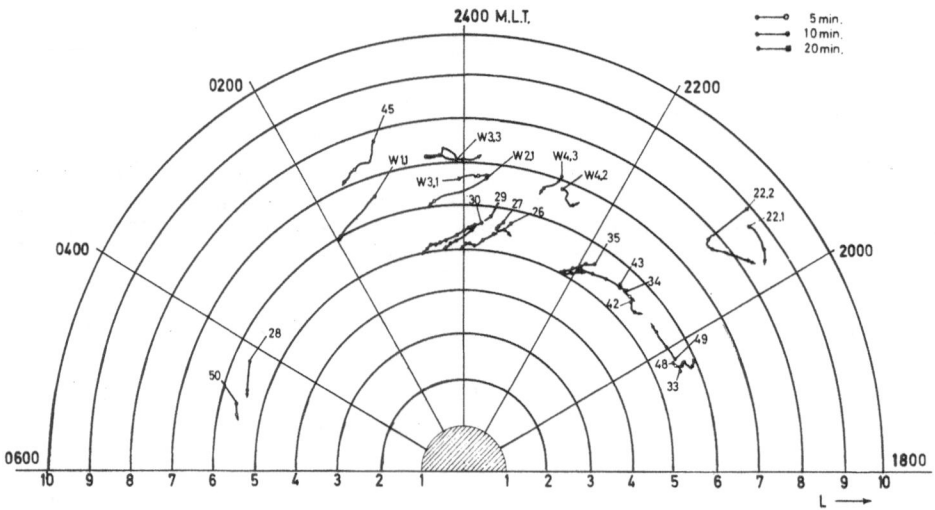

Fig. 9. Data from Figure 8 in an L versus MLT plot.

by a W) in a polar plot versus magnetic local time (MLT) is given in Figure 8. Low ionospheric fields (i.e. corotation) appear as a smooth drift towards dawn. This plot uses the invariant latitude instead of the geomagnetic latitude that was chosen by Haerendel and Lüst (1970) in an earlier presentation of these data. Slight changes are the result. One main observation is that drifts towards dawn (i.e. to the east) are generally accompanied by southward components, whereas strong drifts towards dusk tend to have slight northward components. During quiet periods even at $L=6$ corotation is a typical feature.

A *projection* of these drifts into the magnetic equatorial plane is quite illustrative. If the magnetic field can be considered as frozen-in and is well represented by the

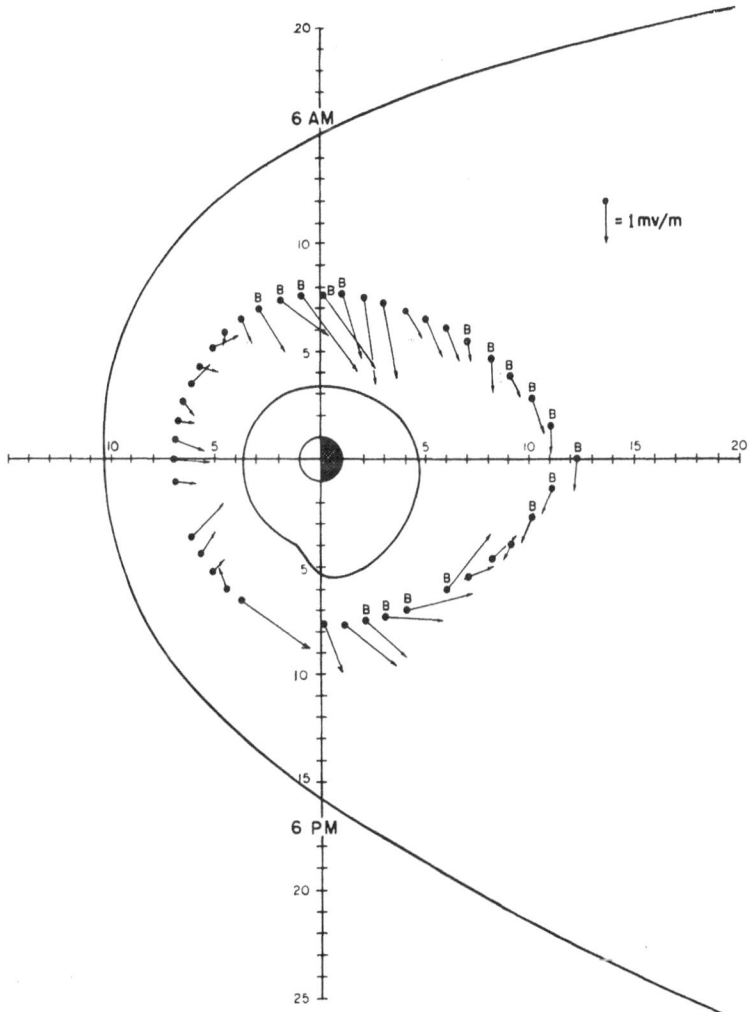

Fig. 10.   Electric field vectors derived from double probe measurements with balloons and projected into the magnetic equatorial plane using Fairfield's (1968) model of the distorted geomagnetic field (Mozer and Serlin, 1969). Letter B indicates the presence of magnetic bays.

model used, we should thus get an impression of the plasma convection in the magnetosphere (Figure 9). At high latitudes L is no longer a good measure for the equatorial distance. This was taken into account by Mozer and Serlin (1969) who projected the electric field derived from balloon data into the magnetosphere by using Fairfield's (1968) model of the magnetic field (Figure 10). The qualitative picture arising from the latter data representing 24 hr of balloon flight is in agreement with the barium cloud data (Figure 9). The matter of the electric field mapping was discussed by Haerendel and Lüst (1970) and Mozer (1970).

Very recently data from *the polar cap* became available. Double probes flown on the polar satellites OGO 6 (Aggson *et al.*, 1970) and Injun 5 (Cauffman and Gurnett, 1970) show that north of the auroral oval the prevailing drifts reverse again. They are directed essentially eastward in the evening and westward in the morning. Magnitudes up to 300 mV/m were reported (Cauffman and Gurnett, 1970). Two weeks of OGO 6 measurements along the dawn-dusk meridian give the impression of a rather homogeneous electric field pointing from dawn to dusk of typically 30–50 mV/m over the polar cap. In the Injun 5 data the field strengths tend to decrease at latitudes above 80°.

Ion cloud experiments of Wescott *et al.* (1970) on the polar cap confirm the observation of eastward drifts in the evening and westward drifts in the morning and rather smooth fields of 15–45 mV/m. One of these experiments was included in Figure 8. It is interesting to note that the close correlation of the directions of $E_\perp$ and of the magnetic perturbation vectors on the ground as typical for the auroral zone is not maintained on the polar cap (Section 3.2).

2.3 RELATION TO AURORAL ARCS

Of much interest is the relative orientation of the electric field vectors and auroral arcs.

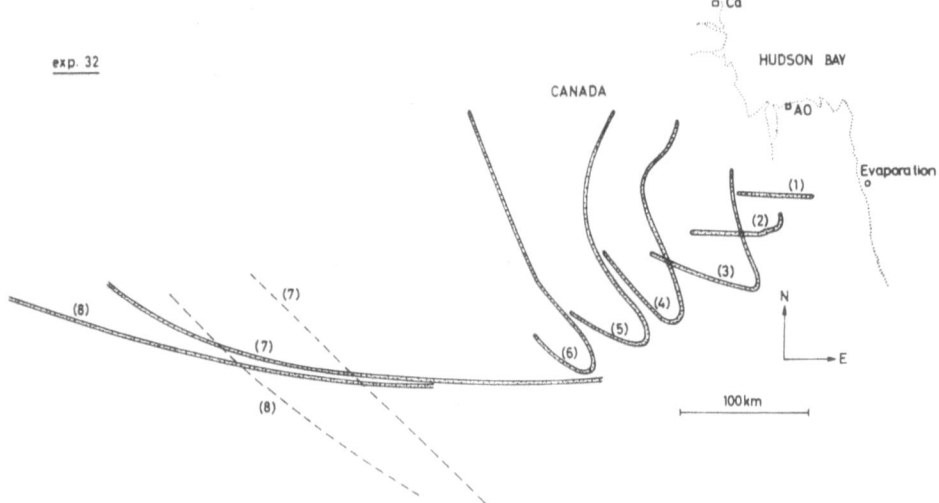

Fig. 11.   Example of the motion of a strongly distorted ion cloud (dotted bands) at 260 km with respect to a later appearing auroral arc (dashed line) (Haerendel *et al.*, 1969). The consecutive times are 370, 555, 755, 965, 1145, 1385, 2135, and 2900 sec after launch.

Wescott *et al.* (1969) found that in their barium cloud experiments the electric field was very nearly normal to the arcs. Haerendel *et al.* (1969) reported an example where the auroral arc intersected the path of the rather distorted ion cloud, but participated in its westward drift (Figure 11). A similar observation was made in another experiment (not yet published). Although it is often observed that motions of auroral forms are in accordance with ion cloud drifts or with the otherwise measured electric field (Kelley *et al.*, 1970), there are cases of disagreement. Wescott *et al.* (1970) observed a poleward moving auroral arc crossing the field lines intersecting an ion cloud. During the time of contact the electric field was considerably depressed (Figure 12). Similar observations were made by Kelley *et al.* (1970) who found generally a close

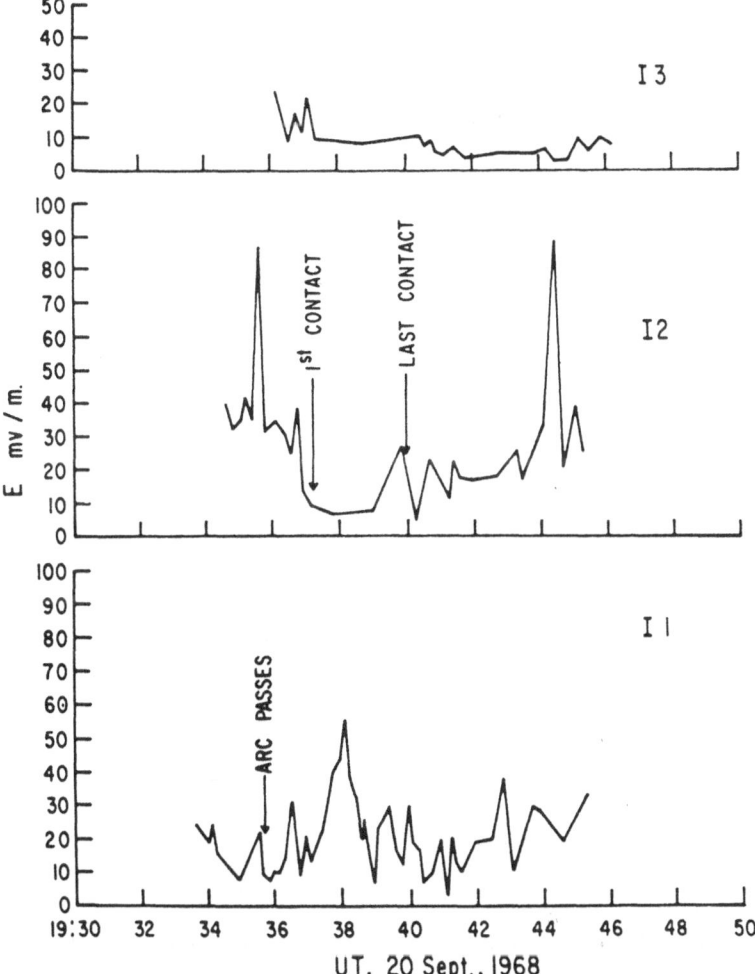

Fig. 12. Electric fields pointing essentially northwestward as derived from the motion of three barium clouds (Wescott *et al.*, 1970). When a northward moving auroral arc crosses the path of the ion clouds the electric field is depressed. The third cloud moved to the southeast and stayed later in contact with the arc.

correlation between north-south motions of auroral forms and electric fields derived from balloon data, but during the onset of magnetic bays near midnight they detected characteristic deviations. These observations either indicate that the motions observed in auroral forms are not all the time due to physical motions of the source particles, or that the concept of frozen-in magnetic field lines is not always applicable over their full extension in the magnetosphere. The depression of the electric field inside auroral arcs was also observed by Aggson (1969) (Table II). Whether this is a general feature is not yet clear.

TABLE II

Some double probe measurements of auroral electric fields (Aggson, 1969)

Auroral form	Magnetic activity	$\Delta X [\gamma]$	Electric field strength [mV/m]	
			in auroral form	out of auroral form
Quiet arc	Quiet	$+20$	<5	20–40
Unknown	Active	-300	–	10–30
Rayed arc	Active	-150	<10	10–100
Weak arc	Quiet	-20	–	20–30
Folded arc	Active	$+10$	<7	–
Folded arc	Active	-130	<4	10–20

2.4. MEASUREMENTS IN THE MAGNETOSPHERE

Measurements of electric fields at high altitudes are scarce. The most reliable data were obtained by the tracing of whistler ducts (method 2.1.3) which yields only data on the radial drift components. One such measurement is shown in Figure 13 (Carpenter and Stone, 1967). The plasmapause starts to contract half an hour before the onset of the substorm at the $L=7$ stations. The inward speed was about 0.4 R_e/hr corresponding to a westward field of 0.3 mV/m at the equator. Other examples of radial motions were reported by Carpenter on several occasions. The correlation of such events with other phenomena like IPDP is of considerable interest (Carpenter *et al.*, 1970).

A direct measurement of the motion of the magnetospheric plasma was achieved by Freeman (1968, 1969) and Freeman *et al.* (1968) with a retarding potential analyzer on ATS 1 (method 2.2.1). Anisotropic velocity distributions of ions in the energy range from 0–50 eV were found on a few occasions during magnetic storms. The results of one event during the storm of January 13–14, 1967, are summarized in Figure 14. In the noon to dusk sector the flow was essentially directed towards the sun. The electric field corresponding to the particle energies via Equation (2) was derived as 5 mV/m. This value is rather uncertain because of the unknown probe potential. When the magnetopause crossed the position of the satellite, the flow of ions was observed to be directed away from the sun and essentially parallel to the magnetopause outside as well as for some distance inside the boundary (Figure 15). This suggests that the magnetopause is an electric equipotential surface and that there is a downwind flow inside the magnetopause similar to that postulated by Axford and Hines (1961).

An extension of the barium cloud technique to high altitudes was attempted in a release from the satellite HEOS 1 at 12.5 R_e on the morning side of the magnetosphere. The response of the artificial plasma to the ambient electric field in the tenuous plasma of the magnetosphere turned out to be fundamentally different from that in the ionosphere (Haerendel and Lüst, 1970). At high altitudes, in particular in the magnetic tail outside the plasmasheet as in the case of the HEOS experiment, the barium plasma is too heavy a load on the magnetic field lines to attain quickly the convective velocity of the ambient plasma. For most of the observation time of 25 min, therefore, the barium cloud was not significantly affected by the ambient plasma flow. However, a weaker structure separating from the main body of the cloud could be observed. The motion of this structure allowed to set a lower limit of 0.085 mV/m to the ambient electric field magnitude.

In this context we should mention some of the attempts to derive indirect conclusions on the electric field by calculating energetic particle paths in the magnetosphere under the action of a model field and comparing the theoretical particle distributions with

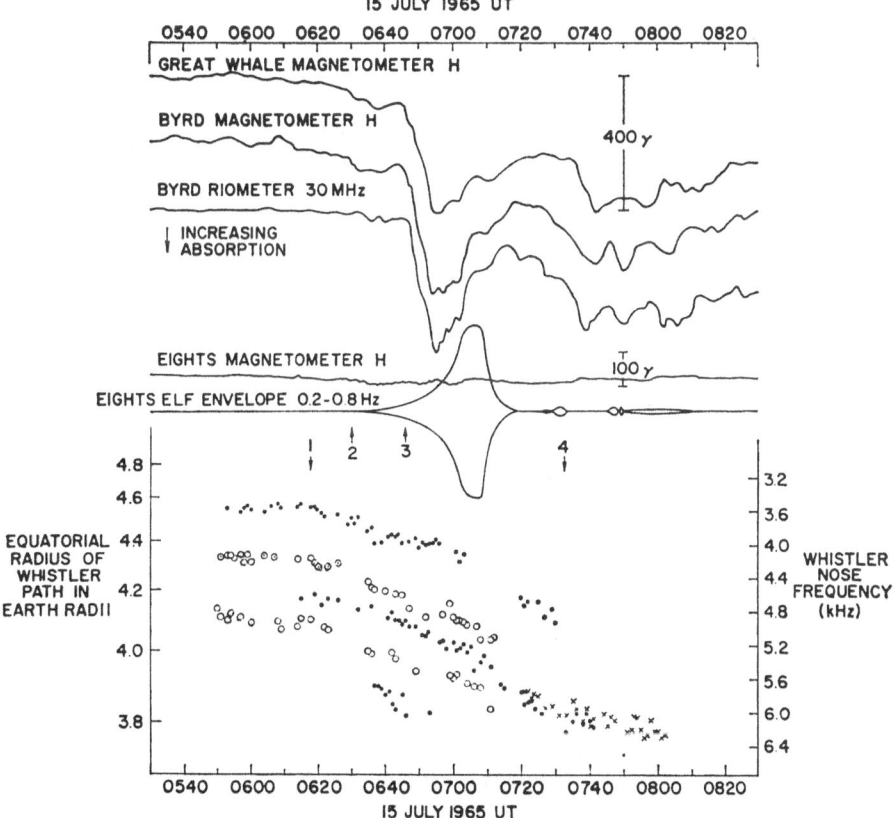

Fig. 13.   Inward motions of several whistler ducts preceding and during a substorm in the midnight sector (Carpenter and Stone, 1967).

observations (Taylor and Hones, 1965; Taylor, 1966). More recently studies of the local time asymmetries of trapped electrons (McDiarmid *et al.*, 1969; Roederer and Hones, 1970), of sudden changes of the particle fluxes at the geostationary orbit during substorms (Winckler, 1970), and of the shadowing effect of the moon on solar electrons (Van Allen and Ness, 1969) have added various estimates of the electric field. McIlwain and de Forest (1970) used the energy dispersion of arrival times of plasma clouds at the ATS V satellite to test various assumptions on the electric fields. General con-

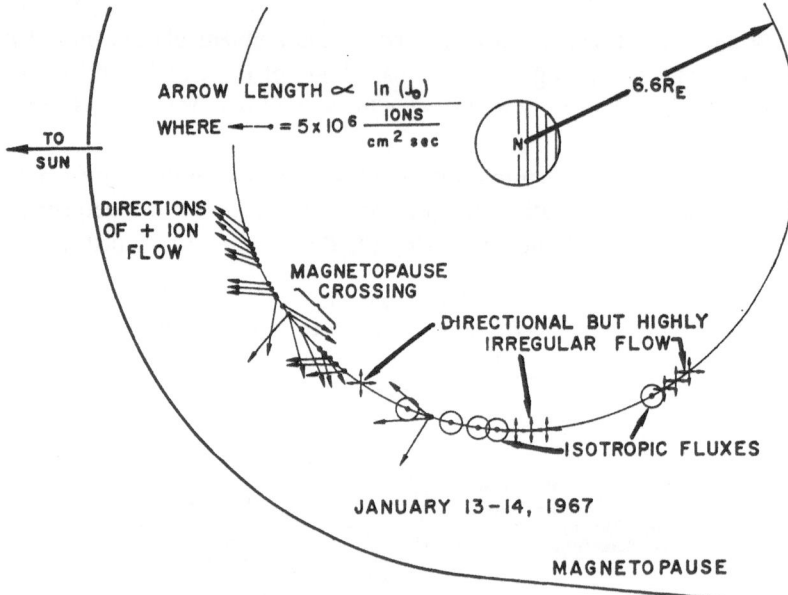

Fig. 14. Plasma flow at the geostationary orbit measured with the retarding potential analyzer on ATS 1 during the magnetic storm of January 13–14, 1967 (Freeman, 1969). When the magnetopause approaches and crosses the satellite the otherwise sunward directed flow reverses its direction.

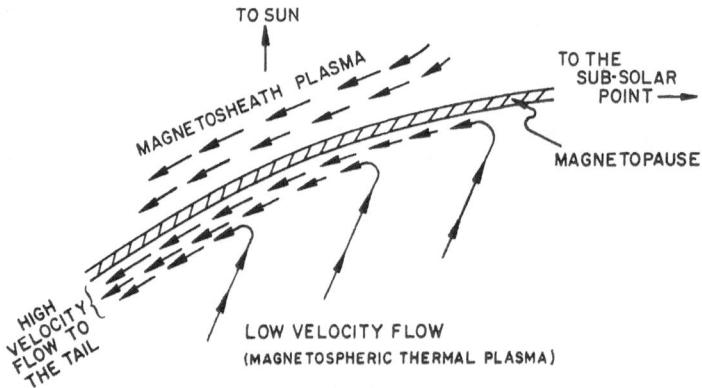

Fig. 15. Inferences on the magnetospheric plasma convection near the magnetopause derived from the measurements on ATS 1 at the magnetopause crossing shown in Figure 14 (Freeman *et al.*, 1968).

TABLE III

Measurements and estimates of magnetospheric electric fields

Method	Author	Typical field strengths [mV/m]	Orientation	At a distance of	Remark
Whistler ducts	Carpenter and Stone (1967)	0.3	Westward	4–5 R_e	Preceding and during bay close to midnight
Electrostatic analyzer	Freeman (1969)	5	Dawn to dusk	6.6 R_e	Magnetic storm
Derived from bay current model, compared with energetic particle distributions	Taylor and Hones (1965)	2	Mostly radially inward	5–7 R_e	Substorm conditions
Local time asymmetry of > 35 keV electrons related to simple electric field models	McDiarmid et al. (1969) Roederer and Hones (1970)	0.06	Dawn to dusk Dawn to dusk	Tail Tail	Topology of Brice's (1967) model used
Characteristic changes of electron fluxes	Winckler (1970)	4	Dawn to dusk	6.6 R_e	Applies to rapid inward motion during substorms near and after midnight
Motion of structure in HEOS barium cloud	Haerendel and Lüst (1970)	> 0.08	Between eastward and northward	12.5 R_e	Experiment in the tail outside the plasma sheet
Shadowing of energetic electrons by moon	Van Allen and Ness (1969)	< 0.5	–	Orbit of moon	–
Dispersion of energetic plasma clouds	McIlwain and de Forest (1970)	0.15	Radial	6.6 R_e	No answer on azimuthal component of E

sistency with the pattern proposed by Axford and Hines (1961) was found. Field strengths on the order of 0.1 mV/m were sufficient to explain the observations. The results of all these estimates are summarized in Table III. They do not lead to a very clear picture.

3. Electromagnetic Coupling of Ionosphere and Magnetosphere

The existing data are far from giving a complete picture of the electric field in the magnetosphere and ionosphere. However, several conclusions can already be drawn. In this chapter only a few features shall be discussed that throw some light on the electromagnetic coupling of the ionosphere and magnetosphere.

3.1. ATMOSPHERIC DYNAMO AND MOTOR

The characteristic difference of the electric fields at medium and low latitudes from those at high latitudes reflects strongly their different origins. It is very likely that most of the quiet time motions inside the plasmasphere (relative to pure corotation) are caused by the dynamo effect of the upper atmosphere. (Radial motions at higher L values during magnetic bays as observed by Carpenter (Section 2.4) and possibly a strong 'viscous' interaction with faster moving plasma just outside the plasmapause are different matters.) The resulting electric polarization fields were found to be generally opposite to the direction of the Sq current during twilight hours except at the equator. These directions are such that there is a steady, although rather slow transport of ionization from the sunlit to the dark hemisphere with speeds of typically 50–100 m/sec. The overall effect of this transport on the plasma density distribution in the ionosphere and plasmasphere has still to be investigated. According to Wagner (1969) the quiet time electric field even at auroral latitudes can be attributed to the Sq-current system.

The absence of strong electric fields at medium latitudes during major magnetic perturbations (Section 2.1) suggests that the horizontal return current from the polar electrojet in the low latitude ionosphere is considerably smaller than the equivalent current. There is, however, another process linking low latitudes to the events in the auroral ionosphere that is ultimately caused by electric fields. It is provided by the acceleration of the neutral atmosphere in the auroral F layer by ion drag as well as by horizontal pressure gradients caused by the Ohmic heating in the auroral oval. On several occasions strong deviations of the neutral wind from the normal pattern were observed during strong magnetic perturbations. Figure 16 shows as an example the drift paths of an ionized and two neutral artificial gas clouds of an experiment carried out towards the end of a four-hour positive magnetic perturbation of the order of 200 γ. Neutral wind speeds of up to 500 m/sec in the general direction of the ion drift were observed. Rees (1970) established a good correlation with the average magnetic perturbation field during two hours preceding the observation time. Two hours is the approximate time constant for the acceleration of the neutral air in the F layer.

This motion must propagate to lower latitudes. Viscous damping is rather inefficient. The dominant deceleration process is again ion drag in the region of slow convection (Kohl and King, 1967). The barium cloud experiments of AFCRL during $K_p = 7+$, which were mentioned in Section 2.1, showed rather weak electric fields, but unusually strong south-westward directed neutral winds of 210 to 265 m/sec at 42° magnetic latitude during sunset (Eglin/Florida). It is tempting to assume that this wind was set up by ion drag on the evening side of the auroral zone where unusually strong positive bay events during the preceding hours indicated the existence of strong westward motions.

King *et al.* (1967) showed that the regular daily variations of the F 2-layer peak can be essentially related to the global neutral wind pattern. It may well be that the anomalous behaviour of the F layer at medium latitudes during and after stronger magnetic bays can be explained by a disturbed neutral wind at F-layer altitudes due to ion drag and additional heating in the auroral oval.

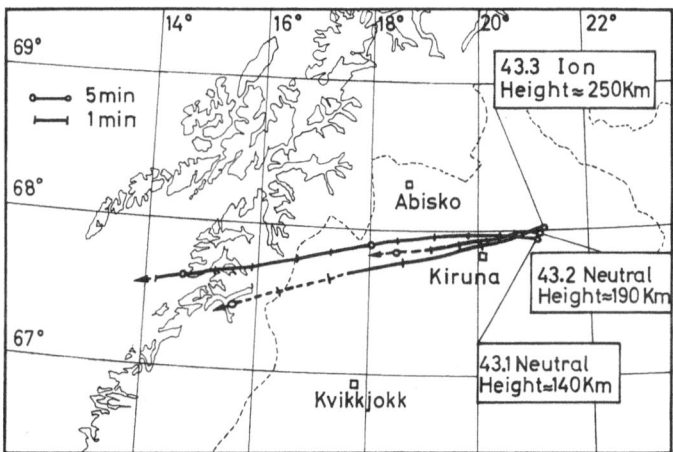

Fig. 16. Paths of one ionized and two neutral gas clouds during a strong positive bay event on March 17, 1969 in Kiruna demonstrating the effects of the ion drag on the neutral atmosphere. The ions move about twice as fast as the neutrals, whose velocity is about 450 m/sec.

3.2. IMPLICATIONS ON ELECTRIC CURRENTS

In evaluating magnetic perturbations on the ground in terms of electric fields in the ionosphere it is an essential question to know the relative contributions of the Hall, Pedersen and possibly field-aligned currents. Föppl *et al.* (1968) derived integrated conductivities from electron-density measurements yielding ratios of Hall over Pedersen conductivity close to two. Combined with the simultaneously measured electric field and under the assumption of a homogeneous entirely horizontal current, good agreement of derived and measured magnetic perturbation fields was obtained. Wescott *et al.* (1969) concluded from a comparison of barium cloud motions with the magnetic perturbation that in the auroral zone the ionospheric current is predominantly a Hallcurrent.

With the presently available material on auroral electric fields this question can be tested more closely. In Figure 17 the azimuth, δ_E, of $\mathbf{E}_\perp$ (measured in a plane normal to $\mathbf{B}$) and the azimuth, δ_H, of the horizontal magnetic perturbation vector, $\Delta\mathbf{H}$, are plotted for a number of barium cloud experiments over northern Scandinavia (measurements of the Garching and Goddard groups). Only those experiments were selected where $\Delta\mathbf{H}$ was sufficiently strong to derive δ_H reliably. That means that the results of this study are typical for substorm conditions. Obviously there is a strong correlation between the directions of $\mathbf{E}_\perp$ and $\Delta\mathbf{H}$. The diagonal lines in Figure 17 show

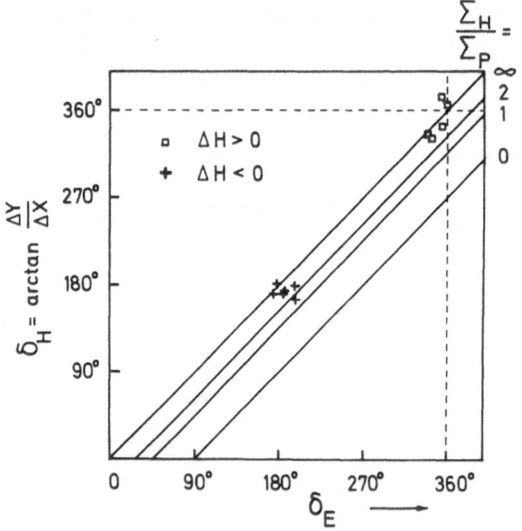

Fig. 17.   Correlation between the azimuth, δ_H, of the horizontal magnetic perturbation vector on the ground and that, δ_E, of the transverse electric field in the ionosphere. δ_E is measured in a plane normal to B and positive from north to east. Data from barium cloud experiments in northern Scandinavia during magnetic perturbations with $|\Delta H| > 40\ \gamma$. Diagonal lines indicate the expected relations for homogeneous horizontal currents with various ratios of integrated Hall and Pedersen conductivities (Σ_H/Σ_p).

the relations expected for a purely horizontal current with different ratios of Hall and Pedersen conductivities. The data group quite clearly on rather large values of this ratio, larger than one would expect theoretically on the grounds of measured or otherwise derived electron-density profiles (Boström, 1964; Föppl *et al.*, 1968; Rees and Walker, 1968).

A way out of this discrepancy is to abandon the assumption of a purely horizontal current, what in the case of an externally applied electric field cannot be avoided (Vasyliunas, 1968). If the Pedersen current, which is directed essentially north-southward (Figure 17), closes predominantly via magnetic field-aligned currents to the source region of the electric field in the magnetosphere, an observer on the ground would see very little effect of this current component. The reason is that the north-south dimension of this current loop should, similarly to auroral forms, be much less than the east-west dimension. Such a topology of currents was suggested by Boström (1964) in one of his models for the auroral electrojet. It is indicated schematically in Figure 18a.

Figure 19, on the other hand, shows the poor correlation of the magnitude of the magnetic perturbation fields, ΔH, with that of $E_\perp$. This demonstrates the strongly variable ionization of the lower ionosphere at high latitudes. We conclude that the magnetic perturbations on the ground are good indicators of the direction, but not of the magnitude of the field. However, this applies only to the auroral zone.

Contrary to the auroral zone, strong disagreement was found on the polar cap between the electric field orientation and that of the magnetic perturbation vector, in the sense that the latter cannot be fully attributed to an overhead ionospheric current driven by the measured field (Wescott et al., 1970; Haerendel and Lüst, 1970).

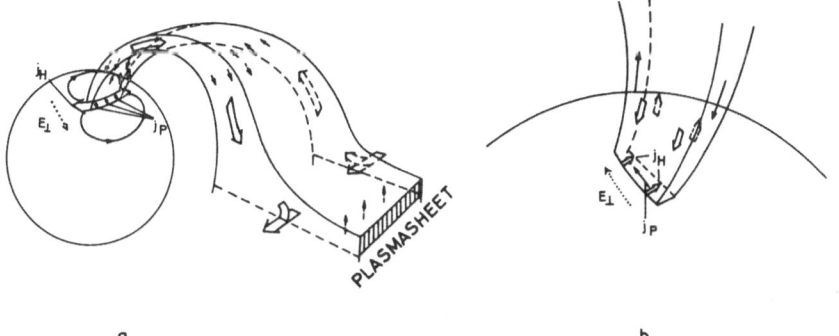

a   b

Fig. 18.   (a) Model of the current distribution in the westward electrojet/plasmasheet current system. The Hall current continues the tail and partial ring currents. The momentum of the easterly drifts dissipated in the ionosphere by the Pedersen current is replenished by the corresponding magnetospheric closure currents. Only part of the electrojet closes horizontally in the ionosphere.
(b) Possible, but incomplete current distribution in auroral forms where $E_\perp$ has a tangential component. The Hall current transverse to the arc can close partly via field aligned currents.

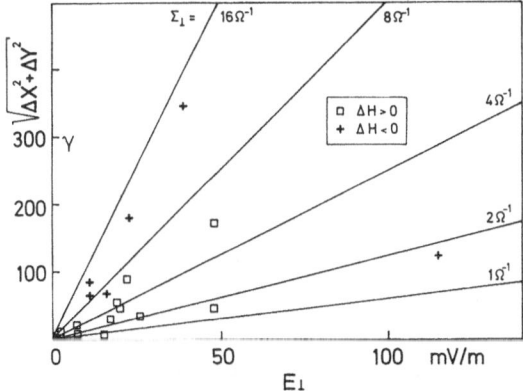

Fig. 19.   Observed relations between horizontal magnetic perturbations on the ground and $E_\perp$. Straight lines indicate the expected behavior for a homogeneous horizontal current layer with a fixed value of the total transverse conductivity.

The discrepancy can be resolved by assuming that the perturbation field a few degrees north of the auroral oval is essentially caused by field aligned currents entering the oval on the morning side and leaving it on the evening side and thereby connecting the auroral electrojet to a magnetospheric current, as suggested by Akasofu and Meng (1969) (see also Akasofu (1970) and Vasyliunas (1970)). Recently Akasofu. *et al.* (1970) presented substorm magnetic field data on the earth's surface and in the magnetic tail supporting this hypothesis.

If we combine this evidence with the previously discussed one on the magnetospheric connection of the Pedersen current we get the following model of the currents and fields maintaining the auroral electrojet: The electrojet is essentially a Hall current driven by a poloidal electric field. Because of the curl-freeness of the electric field it closes partially in the ionosphere, but a significant, if not dominant, fraction is connected to a current system in the plasma sheet via field-aligned currents. It may well be that the westward electrojet continues the near earth part of the tail current on the morning side, and is on the evening side continued by the partial ring current measured by Cummings and Coleman (1968) and Frank (1969). The Hall current is dissipation-free. The dissipating part, the Pedersen current, is essentially normal to the east-westward orientation of the electrojet. Via magnetic field-aligned sheet currents it is connected to the region in the magnetosphere where energy and momentum dissipated in the ionosphere are replenished (Walbridge, 1967) (see Figure 18a).

This model, suggested by electric field data, is not the only conceivable one. In particular, inside auroral forms other configurations are possible. As discussed in Section 2.3 the electric field is not necessarily normal to auroral forms. As the tangential component of **E** must be continuous through the surface of an arc we conclude that in these cases there must be a Pedersen current parallel to the arc. Inside the arc the conductivity is strongly enhanced. Therefore, an enhanced Hall current is flowing normal to the arc. This may be partly counteracted by a Pedersen current driven by a secondary normal polarization field (Boström, 1964) or another part may continue via field-aligned sheet currents to the magnetosphere. Magnetic field measurements from a rocket firing through a rather quiet auroral arc by Cloutier *et al.* (1970) are consistent with such a model, of which a simplified version is sketched in Figure 18b.

In summary we can say that the presently available material on the electric field in the ionosphere can merely give hints on such fundamental problems as configuration, continuation, and maintenance of ionospheric current systems. Much more material has to be collected, in particular in the magnetosphere, and be related to simultaneous measurements of the magnetic field and of the conductivity distribution in order to successfully approach final clarification.

4. Summary and Conclusions

In summarizing the main topics of the preceding review of data on d.c. electric fields we shall add a few remarks on magnetospheric convection theories (an excellent review

was given by Axford (1969)). The data do not yet allow firm conclusions. Owing to the relatively small quantity of observational results there is a danger of overemphasizing the significance of one or the other, or on the other hand of disregarding an important feature. The main problem, however, is the incompleteness of coverage in local time and spatial distribution, in particular in the magnetosphere proper.

The ionospheric electric field at low and medium magnetic latitudes has typically a magnitude of 1–2 mV/m, except inside the equatorial electrojet where a vertical polarization field exists that is stronger by one order of magnitude. The data available seem to support the dynamo theory for the Sq-current system. During twilight hours the electric field tends to oppose the direction of the Sq current indicating the relevance of the dynamo fields at these local times or possibly some contribution of magnetospheric currents to the magnetic perturbation field. The plasma drifts resulting from the twilight electric field are directed overwhelmingly from the sunlit to the dark hemisphere.

Except for the radial inward drift of whistler ducts just inside the plasmapause preceding and during a substorm no enhancement of the electric field at medium and low latitudes during major polar disturbances was found by direct measurements. Furthermore, no indication of the electric fields that are expected from Nishida's (Nishida et al., 1966; Nishida, 1968a, b) theory of the DP 2-current system was found. Although DP 2 should be a rather common phenomenon, these conditions may just have been missed. During twilight the expected field should essentially oppose the so far measured one (Figure 4) and should be about an order of magnitude higher in view of the relatively low integrated conductivities (1 Ω^{-1} or less) during night and twilight hours.

The existing data seem to imply that theories of low latitude return currents should be checked for possible contributions of magnetospheric, in particular field-aligned currents to the equivalent current systems (Bramley and Pudovkin, 1972). Low latitude F-layer effects associated with polar magnetic disturbances may be related to changes in the neutral wind pattern set up in the auroral oval.

At high latitudes a significant difference between auroral oval and polar cap was found with regard to the relation between electric fields and equivalent currents. Whereas in the auroral oval the magnetic perturbations (whenever they are sufficiently strong) can be attributed to an ionospheric Hall-current, this does not apply on the polar cap. The attempts to derive the magnetospheric convection pattern from equivalent current systems are therefore subject to errors (Axford and Hines, 1961; Taylor and Hones, 1965; Nishida, 1966; Obayashi and Nishida, 1968; Heppner, 1969). At least during substorms a significant contribution to the magnetic perturbations on the polar cap seems to be of magnetospheric origin, in particular from field-aligned currents. In the auroral zone, the inclusion of the Pedersen conductivity in converting from equivalent currents to ionospheric drift patterns (Taylor and Hones, 1965) is misleading because the Pedersen current seems to connect quite often and substantially to the magnetosphere (Walbridge, 1967; Vasyliunas, 1968).

Although the information obtained from electric field measurements deviates in

some details from convection models due to these circumstances the data so far tend
to support the essential features of these theories initiated by Axford and Hines (1961).
This is partly due to the usual neglect of the Pedersen currents in most of the investigations and, as far as the polar cap is concerned, the fact that not only equivalent
currents, but also other informations, like the alignment of auroral arcs, and physical
intuition were used in the argumentation.

The rather scattered data on the polar cap are consistent with a general flow in
the antisolar direction towards the midnight region. Typical field strengths of several
tens of mV/m and a rather smooth horizontal distribution were found. Unfortunately,
no measurements just a few degrees north of the auroral oval near midnight are
reported. If the one measurement of an antisolar pointing flow close to the magnetopause (Figure 15) is characteristic, this would be the magnetospheric counterpart of
the antisolar pointing drifts in the polar ionosphere. These motions would then have
to be regarded as the driving part of the convection system, set up by 'friction' with
the solar wind (Dungey, 1961; Axford and Hines, 1961; Levy *et al.*, 1964; Axford,
1964). During substorms, however, the release of internal stresses built up by the
quiet time convection may be the more immediate cause (Axford, 1967, 1969).

The drift pattern in the auroral oval is consistent with its interpretation as a return
flow towards dawn and dusk (Axford and Hines, 1961) with a separation line typically
one hour before magnetic midnight (Heppner, 1969). The fields are extremely variable
in magnitude and direction. Field strengths range from a few to a few hundred mV/m
with the most typical values between 10 and 30 mV/m. Strong predominance of either
essentially northward (evening hours) or essentially southward (post-midnight hours)
directed fields is observed (Figure 17). The eastward flow after midnight is commonly
accompanied by a southward component in magnetic coordinates; so the flow is not
parallel to the auroral oval (Figure 8). Such a behavior was reproduced by Vasyliunas
(1970) in his treatment of a polar current system that takes into account an auroral
belt of high ionospheric conductivity and field-aligned currents. In addition, the
rotation of the equivalent current with respect to the Hall current on the polar cap
is also a result of this model. However, the strong westward drifts and the tendency
of slight northward motions observed in the dusk to midnight sector do not show up
clearly in Vasyliunas' theory.

This general pattern must not be thought of as a smooth flow. In particular during
substorms reversals of the flow directions within a few minutes were found with ion
clouds (Figure 8). Often parallel sheets of opposing east-west drifts separated by only
a few to a few tens of kilometers are observed. A fine structure of the electric field
with scale sizes ranging from 100 m to several km as appearing in the field-aligned
striations of barium clouds seems to be a typical feature in the auroral zone. These
geometric patterns bear a great resemblance with the structure of the aurora. Separation
of ionospheric and magnetospheric causes of the fine structure was not yet possible.
No simple general picture of the electric field distribution inside auroral arcs could
be derived. Often the arcs are aligned along the direction of plasma drift and the
electric field strength is considerably reduced inside. In other cases auroral arcs appear

inclined with respect to plasma flow lines. Not all auroral motions, in particular in the break-up phase, can be attributed to the electric fields observed in the ionosphere. A wave-like modulation of the source of precipitating particles may be the cause.

So far the role of the plasmapause in the magnetospheric convection pattern is not quite assessed. According to the theories of Nishida (1966) and Brice (1967), the plasmapause separates magnetic flux tubes that are linked, at least part of the time, to the interplanetary field from those entirely confined to the magnetosphere and therefore closed all the time. The plasmapause is considered as an electric equipotential, i.e. a flow line. The existing data do not suffice to decide on the validity of this concept. In a few cases an abrupt transition from low to higher fields was observed, but no diurnal coverage is available. The rather distinct boundary from low and smooth to strong and irregular electric fields found with double probes on polar orbiting satellites (Figure 6) is generally poleward of the plasmapause. In addition, cases were found among the ion cloud experiments where the plasma outside the plasmasphere was quite clearly corotating. It has, however, to be borne in mind that along flow lines linked to the polar circulation cells the convection velocity is not necessarily everywhere very great. Close to the plasmapause, in particular during daytime, the electric field could be of intermediate strength and rather smooth, after most of the energetic particle content has been dumped into the ionosphere. In other words, it is not yet settled whether the boundary between weak and strong electric fields is a flow line of the convection.

This summary shows quite clearly that the existing electric field data are by no means sufficient to finally clarify even one of the most urgent problems of magnetospheric dynamics. Surveys and synoptic measurements of the ionospheric fields at all local times and latitudes, an extension of reliable and sensitive measuring techniques to greater altitudes, high resolution studies in the auroral zone, and ways to measure the rather weak parallel components are the most desirable tasks for the next years. Much of this is being prepared at present.

References

Aggson, T. L.: 1969, in *Atmospheric Emissions* (ed. by B. M. McCormac and A. Omholt), Van Nostrand Reinhold Co., New York, p. 305.

Aggson, T. L. and Heppner, J. P.: 1964, A Proposal for Electric Field Measurements on the Gravity-Gradient ATS-A Satellite, NASA-Goddard Space Flight Center, Greenbelt, Maryland.

Aggson, T. L. and Heppner, J. P.: 1965, A Proposal for Electric Field Measurements on POGO Satellites, NASA-Goddard Space Flight Center, Greenbelt, Maryland.

Aggson, T. L., Maynard, N. C., and Heppner, J. P.: 1970, *Characteristics of Magnetospheric Convection*, presented at the 51. Annual Meeting of the Amer. Geophys. Union, Washington, D.C.

Akasofu, S.-I.: 1970, in *Particles and Fields in the Magnetosphere* (ed. by B. M. McCormac), Reidel Publ. Co., Dordrecht, Holland, p. 34.

Akasofu, S.-I. and Meng, C.-I.: 1969, *J. Geophys. Res.* **74**, 293.

Akasofu, S.-I., Haerendel, G., and Hedgecock, P. C.: 1970, 'A Substorm during the HEOS Barium Cloud Experiment', *EOS Trans. Amer. Geophys. Union* **51**, 403.

Alpert, Ya. L., Gurevich, A. V., and Pitaevskii, L. P.: 1965, *Space Physics with Artificial Satellites*, Consultants Bureau, New York.

Angerami, J. J. and Carpenter, D. L.: 1966, *J. Geophys. Res.* **71**, 711.

Atkinson, W., Lundquist, S., and Fahleson, U. V.: 1969, The Electric Field Existing at Stratospheric Elevations as Determined by Tropospheric and Ionospheric Boundary Conditions, Royal Institute of Technology, Stockholm, No. 69–35.

Axford, W. I.: 1964, *Planetary Space Sci.* **12**, 45.

Axford, W. I.: 1967, in *Aurora and Airglow* (ed. by B. M. McCormac), Reinhold Publ. Corp., New York, p. 499.

Axford, W. I.: 1969, *Rev. Geophys.* **7**, 421.

Axford, W. I. and Hines, C. O.: 1961, *Can. J. Phys.* **39**, 1433.

Balsley, B. B.: 1969, *J. Geophys. Res.* **74**, 2333.

Bering, E., Kelley, M., and Mozer, F. S.: 1970, *EOS Trans. Amer. Geophys. Union* **51**, 404.

Boström, R.: 1964, *J. Geophys. Res.* **69**, 4983.

Bramley, E. N. and Pudovkin, M. I.: 1972, this volume, p. 117.

Brice, N.: 1967, *J. Geophys. Res.* **72**, 5193.

Carpenter, D. L.: 1963, *J. Geophys. Res.* **68**, 1675.

Carpenter, D. L.: 1966, *J. Geophys. Res.* **71**, 693.

Carpenter, D. L. and Stone, K.: 1967, *Planetary Space Sci.* **15**, 385.

Carpenter, D. L., Fraser-Smith, A. C., Unwin, R. S., Hones Jr., E. W., and Heacock, R. R.: 1970. *EOS Trans. Amer. Geophys. Union* **51**, 402.

Cauffman, D. P. and Gurnett, D. A.: 1970, *EOS Trans. Amer. Geophys. Union* **51**, 404.

Chamberlain, J. W.: 1969, *Rev. Geophys.* **7**, 461.

Cloutier, P. A., Anderson, H. R., Park, R. J., Vondrak, R. R., Spiger, R. J., and Sandel, B. R.: 1970, *J. Geophys. Res.* **75**, 2595.

Coroniti, F. V.: 1969, in *Planetary Electrodynamics* (ed. by S. C. Coroniti and J. Hughes), Gordon and Breach Science Publ., Vol. 2, p. 309.

Cummings, W. D. and Coleman, P. J., Jr.: 1968, *Radio Sci.* **3**, 758.

Dungey, J. W.: 1961, *Phys. Rev. Letters* **6**, 47.

Dupree, T. H.: 1970, A Theory of Resistivity in Collisionless Plasma, International Center for Theoretical Physics, Trieste IC/70/40.

Fahleson, U. V.: 1967, *Space Sci. Rev.* **7**, 238.

Fahleson, U. V., Kelley, M. C., and Mozer, F. S.: 1968, Investigation of the Operation of a D.C. Electric Field Detector, Space Sci. Lab., Univ. of Calif., Berkeley.

Fahleson, U. V., Fälthammar, C.-G., Pedersen, A., Knott, K., Brommundt, G., Haerendel, G., and Rieger, E.: 1970, *Simultaneous Electric Field Measurements in the Auroral Ionosphere Using Three Independent Techniques*, presented at the Upper Atmospheric Currents and Electric Fields Symposium, Boulder.

Fairfield, D. H.: 1968, *J. Geophys. Res.* **73**, 7329.

Föppl, H., Haerendel, G., Haser, L., Lüst, R., Melzner, F., Meyer, F., Neuss, H., Rabben, H.-H. Rieger, E., Stöcker, J., and Stoffregen, W.: 1968, *J. Geophys. Res.* **73**, 21.

Frank, L. A.: 1969, *EOS Trans. Amer. Geophys. Union* **50**, 284.

Freeman, Jr., J. W.: 1968, *J. Geophys. Res.* **73**, 4151.

Freeman, Jr., J. W.: 1969, *Science* **163**, 1061.

Freeman, Jr., J. W., Warren, C. S., and Maguire, J. J.: 1968, *J. Geophys. Res.* **73**, 5719.

Gdalevich, G. L.: 1964, *Artificial Earth Satellites* **17**, 43.

Grard, R. J. L., and Tunaley, J. K. E.: 1970, The Photoelectron Sheath around a Body in Interplanetary Space, ESRO Scientific Note, SN-108.

Gurevich, A. V.: 1964, *Cosmic Res.* **2**, 196.

Gurnett, D. A.: 1970, in *Particles and Fields in the Magnetosphere* (ed. by B. M. McCormac), Reidel Publ. Co., Dordrecht, Holland, p. 239.

Haerendel, G. and Lüst, R.: 1968a, in *Earth's Particles and Fields* (ed. by B. M. McCormac), Reinhold Book Corp., New York, p. 271.

Haerendel, G. and Lüst, R.: 1968b, *Sci. Am.* **219**, 80.

Haerendel, G. and Lüst, R.: 1970, in *Particles and Fields in the Magnetosphere* (ed. by B. M. McCormac), Reidel Publ. Co., Dordrecht, Holland, p. 213.

Haerendel, G., Lüst, R., and Rieger, E.: 1967, *Planetary Space Sci.* **15**, 1.

Haerendel, G., Lüst, R., Rieger, E., and Völk, H.: 1969, in *Atmospheric Emissions* (ed. by B. M. McCormac and A. Omholt), Van Nostrand Reinhold Co., New York, p. 293.

Haerendel, G., Lüst, R., Melzner, F., Neuss, H., Rieger, E., Rhagava Rao, R., Narayanan, M. S.,

and Bhavsar, P. D.: 1970, Barium Release Experiments near the Magnetic Equator at Thumba, India, MPI-Report.

Heppner, J. P.: 1969, in *Atmospheric Emissions* (ed. by B. M. McCormac and A. Omholt), Van Nostrand Reinhold Co., New York, p. 251.

Imyanitov, I. M., Gdalevich, G. L., and Shvarts, Ya. M.: 1964, *Artificial Earth Satellites*, **17**, 66.

Kavadas, A. and Johnson, D. W.: 1964, in *Space Res.*, Vol. IV (ed. by P. Muller), North-Holland Publ. Co., p. 365.

Kelley, M., Serlin, R., and Starr, J.: 1970, *EOS Trans. Amer. Geophys. Union* **51**, 404.

Kellogg, P. J. and Weed, M.: 1969, in *Planetary Electrodynamics* (ed. by S. C. Coroniti and J. Hughes), Gordon and Breach Science Publ., New York, Vol. 2, p. 431.

King, J. W., Kohl, H., and Pratt, R.: 1967, *J. Atmospheric Terrest. Phys.* **29**, 1529.

Knott, K.: 1970, *Space Res.* **10**, 773.

Kohl, H. and King, J. W.: 1967, *J. Atmospheric Terrest. Phys.* **29**, 1045.

Levy, R. H., Petschek, H. E., and Siscoe, G. L.: 1964, *AIAA J.* **2**, 2065.

Liu, V. C.: 1969, *Space Sci. Rev.* **9**, 423.

Maeda, H.: 1955, *J. Geomag. Geoelectr.* **7**, 121.

Matsushita, S.: 1969, *Radio Sci.* **4**, 771.

Maynard, N. C. and Heppner, J. P.: 1970, in *Particles and Fields in the Magnetosphere* (ed. by B. M. McCormac), Reidel Publ. Co., Dordrecht, Holland, p. 247.

McDiarmid, I. B., Burrows, J. R., and Wilson, M. D.: 1969, *J. Geophys. Res.* **74**, 3554.

McIlwain, C. E. and De Forest, S. E.: 1970, *Plasma Clouds in the Magnetosphere*, presented at the Intern. Symp. on Solar Terr. Phys., Leningrad (abstract).

Melzner, F. and Völk, H.: 1970, *Proposal for an Experiment for the ESRO Geostationary Satellite*, ESRO Proposal, S-329.

Mende, S. B.: 1968, *J. Geophys. Res.* **73**, 991.

Mozer, F. S.: 1970, *Planetary Space Sci.* **18**, 259.

Mozer, F. S. and Bruston, P.: 1967, *J. Geophys. Res.* **72**, 1109.

Mozer, F. S. and Serlin, R.: 1969, *J. Geophys. Res.* **74**, 4739.

Mozer, F. S. and Manka, R. H.: 1970, *EOS Trans. Amer. Geophys. Union* **51**, 405.

Nishida, A.: 1966, *J. Geophys. Res.* **71**, 5669.

Nishida, A.: 1968a, *J. Geophys. Res.* **73**, 1795.

Nishida, A.: 1968b, *J. Geophys. Res.* **73**, 5549.

Nishida, A., Iwasaki, N., and Nagata, T.: 1966, *Ann. Geophys.* **22**, 478.

Obayashi, T. and Nishida, A.: 1968, *Space Sci. Rev.* **8**, 3.

Pfister, W.: 1969, in *Small Rocket Instrumentation Techniques*, North-Holland Publ. Co., Amsterdam, p. 66.

Potter, W. E. and Cahill, Jr., L. J.: 1969, *J. Geophys. Res.* **74**, 5159.

Rees, D.: 1970, *Ionospheric Drift Motions in the Auroral Zone during a Substorm*, preprint.

Rees, M. H. and Walker, J. C. G.: 1968, *Ann. Geophys.* **24**, 193.

Rieger, E.: 1969, in *Low Frequency Waves and Irregularities in the Ionosphere* (ed. by N. D'Angelo), Reidel Publ. Co., Dordrecht, Holland, p. 218.

Rieger, E.: 1970, *Measurements of Electric Fields in Equatorial and Mid-Latitudes Using Barium Ion Clouds*, presented at the Intern. Symp. on Solar Regular Daily Geomagnetic Variations, to be published.

Roederer, J. G. and Hones, Jr., E. W.: 1970, *Estimation of the Electric Field in the Magnetosphere as deduced from Trapped Particle Flux Asymmetries*, University of Denver, preprint.

Scholer, M.: 1970, *Planetary Space Sci.* **18**, 977.

Spenner, K.: 1970, *Verhandlungen DPG (VI)* **5**, 647.

Storey, L. R. O., Aubry, M. P., and Meyer, P.: 1968, in *Plasma Waves in Space and in the Laboratory* (ed. by J. O. Thomas and B. J. L. Landmark), University Press, Edinburgh, p. 303.

Swift, D. W.: 1965, *J. Geophys. Res.* **70**, 3061.

Taylor, H. E.: 1966, *J. Geophys. Res.* **71**, 5135.

Taylor, H. E. and Hones, Jr., E. W.: 1965, *J. Geophys. Res.* **70**, 3605.

Taylor, J. C.: 1967, *Planetary Space Sci.* **15**, 155.

Van Allen, J. A. and Ness, N. F.: 1969, *J. Geophys. Res.* **74**, 71.

Vasyliunas, V. M.: 1968, *J. Geophys. Res.* **73**, 5805.

Vasyliunas, V. M.: 1970, in *Particles and Fields in the Magnetosphere*, (ed. by B. M. McCormac), Reidel Publ. Co., Dordrecht, Holland, p. 60.

Völk, H. and Haerendel, G.: 1970, in *Intercorrelated Satellite Observations Related to Solar Events* (ed. by V. Manno and D. E. Page), D. Reidel, Dordrecht, Holland, p. 280.

Wagner, Ch.-U.: 1969, *Some Investigations about Electric Fields at the Basis of the Magnetosphere*, presented at the IAGA General Scientific Assembly, Madrid.

Walbridge, E.: 1967, *J. Geophys. Res.* **72**, 5213.

Wescott, E. M., Stolarik, J. D., and Heppner, J. P.: 1969, *J. Geophys. Res.* **74**, 3469.

Wescott, E. M., Stolarik, J. D., and Heppner, J. P.: 1970, in *Particles and Fields in the Magnetosphere* (ed. by B. M. McCormac), Reidel Publ. Co., Dordrecht, Holland, p. 229.

Whipple, Jr., E. C.: 1970, *Effects of Changing Satellite Potential on Direct Ion Measurements through the Plasmapause*, presented at the Annual Meeting of the American Geophys. Union, Washington, D.C.

Winckler, J. R.: 1970, in *Particles and Fields in the Magnetosphere* (ed. by B. M. McCormac), Reidel Publ. Co., Dordrecht, Holland, p. 332.

Woodman, R. F. and Balsley, B. B.: 1969, *J. Atmospheric Terrest. Phys.* **31**, 865.

Woodman, R. F. and Hagfors, T.: 1969, *J. Geophys. Res.* **74**, 1205.

ELECTRIC FIELDS AND THEIR EFFECTS IN THE IONOSPHERE

(A Theoretical Treatment)

E. N. BRAMLEY

S.R.C. Radio and Space Research Station, Ditton Park, Slough, Bucks, England

and

M. I. PUDOVKIN

Physical Institute, Leningrad State University, Leningrad, U.S.S.R.

Abstract. The sources and magnitudes of electric fields which occur in the ionosphere are briefly described. The relationship between electric fields and currents, and the problems of deducing current systems from magnetic data, are discussed; in particular the DP-current system and its associated electric fields are considered in detail. The effects of electric fields on ionization distributions are described under three headings, namely large-scale distributions, effects on irregular structures, and the production of irregularities.

1. Introduction

The concept of electric currents, and hence the existence of electric fields in the upper atmosphere, is nearly ninety years old (Stewart, 1882); but it is only in the last few years that it has been possible to make reliable direct measurements of such electric fields. Although much remains to be done experimentally in further development and use of the available techniques, it is now possible to make some quantitative estimates of the influence of these fields on the observable properties of the ionosphere.

In this review we first consider briefly the main sources of ionospheric electric fields and then describe in more detail some of the more important effects which they produce. Attention is restricted to electric fields which have periods of the order of hours or more, thus excluding fields associated with micropulsations and ionospheric noise. Unless otherwise stated, the electric field under discussion is that which would be measured by an observer rotating with the earth.

2. Sources and Magnitudes of Ionospheric Electric Fields

2.1. DIFFUSIVE SEPARATION

In the topside of the F region a field arises to counteract the tendency of ions and electrons to form their own diffusive equilibrium distributions under gravity, with different scale heights (Mange, 1960). This field is upwards and is approximately equal to $m_i g/2e$, where e is the proton charge, g the acceleration of gravity, and m_i the mean ion mass. The field is only of order 1 μV/m, but it greatly modifies the distribution of the individual constituents. For instance the scale height of the electron concentration becomes approximately twice that of the dominant ionizable constituent. Also the concentration of a light minor ion such as helium is caused to increase with height in a certain range, thus producing a layer with a maximum

Dyer (ed.), Solar-Terrestrial Physics/1970: Part IV, 117–141. All Rights Reserved.
Copyright © 1972 by D. Reidel Publishing Company.

concentration at a certain height instead of the monotonic decrease which would otherwise occur.

2.2. EARTH'S ROTATION

The earth's rotation is the source of a dynamo electric field which is in the meridian plane and normal to the earth's field lines. This electric field is given by

$$E = (\omega \times r) \times B. \tag{1}$$

It has a vertically upward component which has a maximum value of about 15 mV/m at the equator, and a poleward horizontal component which has a similar maximum value at a latitude of 45°. At low and middle latitudes at least, the co-rotation of the ionosphere with the earth implies that this field is neutralized by a space charge of density

$$\varrho = - \varepsilon_0 \operatorname{div} E = - 2\varepsilon_0 B \cdot \omega. \tag{2}$$

This charge density is zero at a latitude of 35.3°. It is positive towards the poles and negative towards the equator, but the maximum charge density involved amounts to the equivalent of only about 0.32 electrons per cubic metre. For an observer at rest with respect to the sun-earth line, a field $-E$ should of course be measurable; hence when it is required to convert a measured ionospheric field into the equivalent field in the equatorial plane of the magnetosphere, it is usual first to subtract E from the measured value.

2.3. ATMOSPHERIC DYNAMO

Here the driving force is the horizontal movement of neutral air, which in turn arises from thermal and gravitational atmospheric tidal forces. At E-region heights the neutral air wind causes corresponding ionization movements, setting up a $v \times B$ dynamo electric field. The electric currents which are thereby produced have to satisfy the zero-divergence condition, and this involves the establishment of an electrostatic polarization field E_p. The ionospheric currents which are discussed in Section 3 are then determined by the total electric field E and the tensor conductivity σ, the current density J being given by the equation

$$J = \sigma \cdot E = \sigma \cdot (E_p + v \times B). \tag{3}$$

There are two main methods of calculating the electric field. First, one can assume a wind system v and a conductivity σ, and solve Equation (3) subject to the conditions that div $J=0$ and curl $E_p=0$. Various calculations have been made from this approach, using diurnal and/or semidiurnal winds (e.g. Baker and Martyn, 1953; Fejer, 1953; Matsushita, 1953). Secondly, the observed geomagnetic variations at the ground may be used to deduce the ionospheric current system, and hence the electric field and wind system which satisfy Equation (3) (e.g. Maeda, H., 1957, 1963; Kato, 1956, 1957; Hirono et al., 1959).

In the most recent calculations of the Sq and L current systems and electric fields from the dynamo theory (Matsushita, 1969), both methods were combined by selecting an atmospheric tidal system to give the best fit with the observed current system.

Figure 1 shows the Sq electrostatic field distribution so obtained, for equinox months and moderate sunspot conditions. The field is predominantly east–west in the equatorial region, with magnitude of order 1 mV/m, being eastward by day and westward by night. At middle and higher latitudes these zonal components are in the same sense but of somewhat larger magnitude, while there is in addition a north–south component which also amounts to a few millivolts per metre. This component reverses sign in the region of 50°–60° latitude. The corresponding L fields obtained in this work are only of the order of one tenth of the Sq fields.

The importance of these electrostatic fields in the F region, to be discussed in Section 4, depends on their transference into that region from the dynamo levels. Theoretical studies (e.g. Spreiter and Briggs, 1961a, 1961b) show that large-scale fields of the type considered should be very efficiently transferred over this height range.

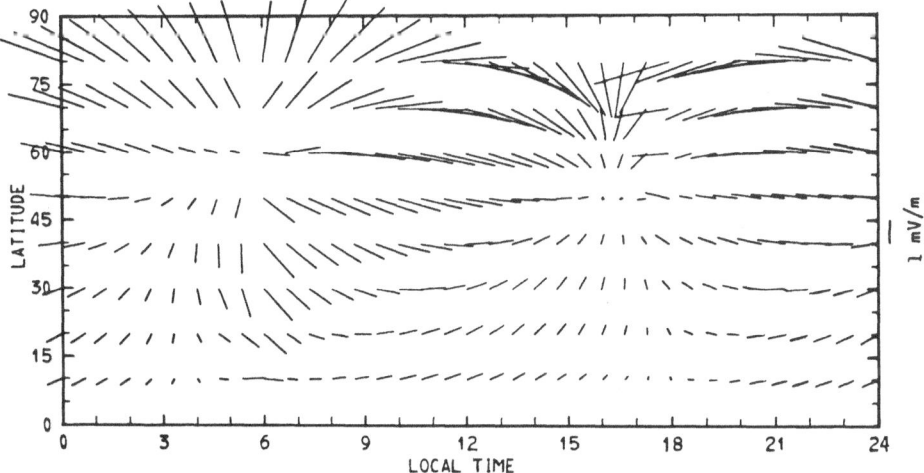

Fig. 1.   Sq electrostatic field distribution (after Matsushita, 1969).

2.4. Magnetospheric dynamo

The final source of ionospheric electric field to be considered is that which is believed to originate in the magnetosphere and which may play an important role in auroral phenomena. The simplest interpretation of this field is as follows. It is supposed that plasma movements v are induced in the distant magnetosphere by interaction with the solar wind, and that this motion sets up an electrostatic field $B \times v$, just cancelling the $v \times B$ field which would otherwise be seen by an observer moving with the plasma. Various mechanisms for the interaction and generation of the field have been considered in detail by different authors (e.g. Piddington, 1960, 1964; Axford and Hines, 1961; Chamberlain, 1961; Dungey, 1963). In each case, the efficient transfer of the electric field into the ionosphere depends on the high conductivity along the magnetic field lines. It is usually accepted that fields with scale size greater than a few tens of kilometres should be transferred into the ionosphere without appreciable attenuation (Reid, 1965), though the possibility of anomalous resistivity along the lines of force, which might invalidate this, has been discussed (Swift, 1965; Ossakow, 1968). In any case, experimental measurements in the auroral regions have definitely established the existence of electric fields much larger than those predicted by the atmospheric dynamo mechanism. It appears that here we are frequently concerned with field strengths up to many tens of millivolts per metre.

3. Electric Fields and Current Systems

The general relation between the current density J and electric field E at any point in the ionosphere has been given in Equation (3), in which the conductivity tensor σ has the following components: σ_0 along the magnetic field; σ_1 (the Pedersen conductivity) parallel to the component of E which is perpendicular to B, and σ_2 (the Hall conductivity) in the direction of $B \times E$. Well-known formulae (Baker and Martyn, 1953) allow the conductivities to be calculated as a function of height, and hence the height-integrated values can be obtained. Table I gives the results of various calculations of Σ_P and Σ_H, the height-integrated Pedersen and Hall conductivities.

TABLE I

Integrated conductivity of the ionosphere

Location	Σ_P (mho)	Σ_H (mho)	Reference
Middle latitudes (day)	6.4	13.6	Baker and Martyn (1953)
	14.4	22.5	Chapman (1956)
	24	33	Maeda and Matsumoto (1962)
	17.6	14.7	Kim and Kim (1962)
	20	26	Brünelli (1963)
	5	10	Fatkullin (1964)
Middle latitudes (night)	1.2	0.47	Maeda and Matsumoto (1962)
	4	0.6	Brünelli (1963)
	0.56	0.19	Boström (1964)
	1.1	0.89	Kim and Kim (1963)
In aurorae	36	56	Boström (1964)
		30	Brünelli (1963)
	214	406	Kim and Kim (1963)
	4.5	10.0 ⎫	
	4.1	8.0 ⎬	Föppl et al. (1968)
	10.1	12.7 ⎭	

It is seen from the table that by day the Hall conductivity is predominant, but the ratio Σ_H/Σ_P hardly exceeds 2. At night the situation is quite different. As the recombination rate rapidly increases with decreasing height, the most drastic changes of the electron density after sunset are observed in the lower ionosphere, where $\sigma_2 \gg \sigma_1$. Therefore at night the integrated Hall conductivity is less than the Pedersen conductivity.

At higher latitudes the height distribution of the electron density is less well known.

Nevertheless it may be supposed that during quiet periods in the winter the conductivity of the polar ionosphere differs insignificantly from that of the middle-latitude ionosphere. Indeed, in the polar night the only source of ionization in the high-latitude ionosphere is corpuscular precipitation, which takes place even on quiet days. However the energy of precipitated particles is not high, and the ionization they produce is located somewhat higher than the Hall layer, which further diminishes the ratio Σ_H/Σ_P.

In the remainder of this section we shall discuss the current system and associated electric fields of a polar geomagnetic disturbance.

3.1. CONFIGURATION OF THE DP-CURRENT SYSTEM

When the distribution of the conductivity in the ionosphere and the configuration of the current system are known, the distribution of the electric field responsible for that current system may be calculated. However at present such a calculation is greatly hindered by the fact that the real current system is not known adequately, in spite of a large number of papers discussing this problem. The very existence of numerous models of the polar storm current system is evidence of such an uncertainty. Some of these current systems will be considered below.

In Figure 2a is shown a classical current system of an averaged magnetic bay by Silsbee and Vestine (1942), and in Figure 2b one of the latest current systems obtained by Afonina and Feldstein (1970), using IGY data. Both systems are constructed assuming that the currents responsible for the magnetic field of a storm flow completely within the ionosphere. The validity of this suggestion is not obvious, and a distribution of the disturbance field quite similar to that observed may be obtained from a fundamentally different distribution of the currents. For example Birkeland (1908) suggested that the currents which are responsible for a magnetic bay flow in the ionosphere only in the region of the auroral electrojet and are closed along the geomagnetic field lines (Figure 2c). Various models of the storm current systems including field-aligned currents of various forms are considered in a number of recent studies. Most of these models can be represented as a superposition (in some proportion) of the Birkeland currents and the currents described by the Boström (1964, 1968) model shown in Figure 2d.

Thus, the data from ground-based observations of the magnetic bay field are insufficient to choose between the various models of the current system, and in particular to decide on the relative importance of field-aligned currents. Additional data are needed, but unfortunately, very few suitable observations are available, and these are sometimes contradictory. The following data are relevant:

(1) The rocket measurements by Cahill (1959) in North Greenland showed the existence of ionospheric currents over the polar cap.

(2) An appreciable increase of the drift velocity of ionization inhomogeneities in the middle-latitude F_2 layer with enhancement of the magnetic activity (Mirkotan, 1962) suggests the appearance or increase of electrostatic fields in the ionosphere at these latitudes. These electrostatic fields seem to be associated with the return currents closing the auroral electrojet.

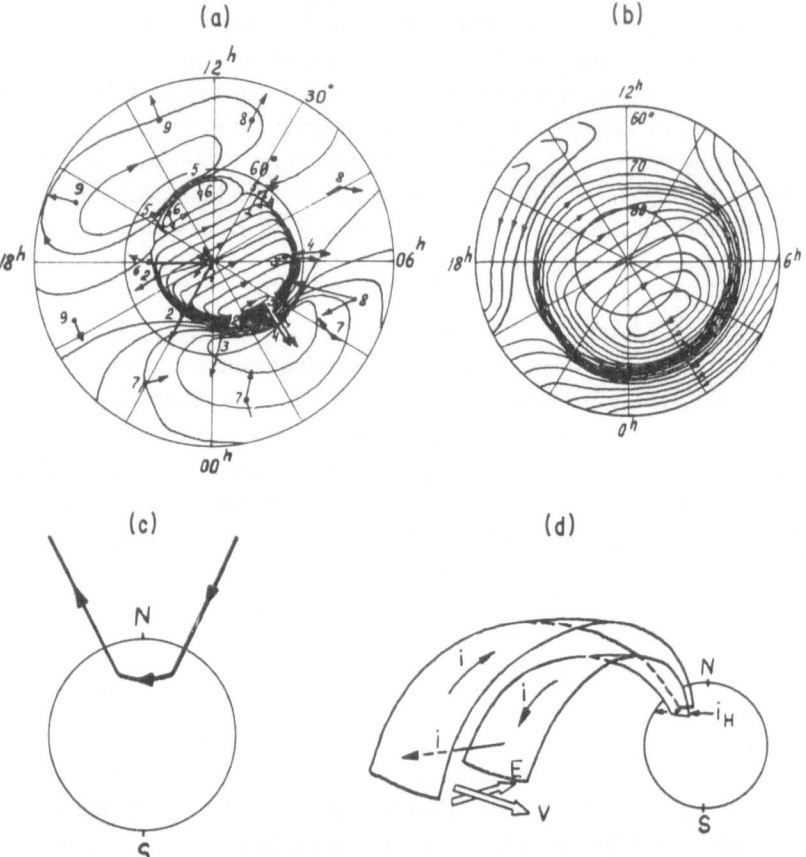

Fig. 2.   Polar storm current systems: (a) after Silsbee and Vestine (1942), (b) after Afonina and
Feldstein (1970), (c) after Birkeland (1908), and (d) after Boström (1968).

(3) During intense magnetic bays there are observed anomalous enhancements of
the disturbance field at the magnetic equator. This also shows that during a storm
an electrostatic field (and, consequently, electric currents) arise in the low- and middle-
latitude ionosphere (Akasofu, 1966).

These data seem to confirm the reality of ionospheric current systems in the polar
caps and at middle latitudes. However there is strong evidence that the ionospheric
current system is not adequate and does not completely represent the real distribution
of the DP-currents. This evidence is:

(1) The data obtained by OGO-2 show that at least in some cases the field of a
magnetic disturbance at low latitudes cannot be explained by ionospheric currents
and is due to the development of an asymmetric ring current in the magnetosphere
(Langel and Cain, 1968).

(2) Zmuda *et al.* (1966, 1967) have shown that during a polar substorm one can
observe relatively intense (up to 100 γ) and very local variations of the geomagnetic
field at a height of about 1000 km, which may be considered as evidence of the

existence of electric currents flowing along the geomagnetic field lines originating near the auroral zone.

Thus, at present we can only state that the auroral currents do close (at least in part) in the ionosphere of the polar caps and at middle latitudes. But the real configuration of the ionospheric current system may differ considerably from the systems shown in Figures 2a and 2b, due to the existence of field-aligned currents.

The intensity and importance of the field-aligned currents are still not known. In order to estimate their possible intensity and influence on the configuration of the ionospheric current system let us consider the model by Boström (1964, 1968) as shown in Figure 2d. Let the auroral electrojet be a Hall current, flowing in a homogeneous and horizontal sheet some hundred kilometres wide (McNish, 1937; Pudovkin, 1960). The linear current density in this sheet is supposed to be j_H(A/m), and the geomagnetic disturbance intensity to be $\delta H(\gamma) = 200\pi j_H$. As there also exists Pedersen conductivity in the ionosphere, a Pedersen current, of linear density $j_P = (\Sigma_P/\Sigma_H) j_H$, must flow across the auroral zone. Supposing that this current closes entirely in the magnetosphere, and that its ionospheric cross-section is a rectangle $1 \text{ m} \times l$ m, where l is the mean width of an auroral arc, the current density of the field-aligned current in the ionosphere is

$$J_{\parallel}^{(i)} = \frac{\Sigma_P}{\Sigma_H} \cdot \frac{\delta H}{200\pi l}.$$ (4)

Hence, taking into account the expansion of the field tube in going away from the earth, by a factor of say 100, the density of the field-aligned current in the magnetosphere is

$$J_{\parallel}^{(m)} = \frac{\Sigma_P}{\Sigma_H} \cdot \frac{\delta H}{2 \times 10^4 \, \pi l}.$$ (5)

The field-aligned current density in the magnetosphere is not known, but it may be assumed that, due to the intensive scattering of particles by ion sound waves, this density cannot exceed the critical value Nev_i, where N is the particle number density in the magnetosphere and v_i is the thermal velocity of the ions (Swift, 1965; Korablev and Rudakov, 1966). Supposing the temperature of the magnetosphere plasma to be 10^4 K (Mayr and Volland, 1968), corresponding to $v_i \approx 10^4$ m/s, and taking into account that the particle concentration hardly exceeds 10^7 m^{-3} outside the plasmapause, the critical density of the field-aligned currents in the magnetosphere proves to be $\approx 2 \times 10^{-8}$ A/m^2. Substituting this value in Equation (5) and supposing $l = 10^4$ m (Kim and Volkman, 1963), with $\Sigma_P/\Sigma_H \approx 0.5$, we obtain a maximum value of δH of order 30γ. This estimate is rather rough, and we may only state that the intensity of the geomagnetic disturbance associated with field-aligned currents has a magnitude somewhere in the range 1–100γ.

Because of this uncertainty in the theoretical data we try to obtain an experimental estimate of the critical intensity of the field-aligned currents in the magnetosphere. According to modern ideas, geomagnetic pulsations are associated with these currents

(Vanyan, 1968). So the development of anomalous resistance of the plasma, limiting the current, may change the character of the pulsations. Figure 3 shows the dependence of the decrement (λ) of high-latitude geomagnetic pulsations (Pi 2 and Pc 3) on the intensity of the magnetic bays (δH), according to Pudovkin and Barsukov (1970). From the figure one can see that the value of λ rapidly increases with increase of δH from 10γ to 100γ. This fact may be interpreted as evidence of the intensive development of the turbulence of the magnetospheric plasma. Thus the experimental data confirm the suggestion that the conductivity of the field lines is sufficient to allow only small disturbances of the geomagnetic field $(\delta H < 10-100\gamma)$. During a polar substorm, the mean intensity of which is several hundred gammas, the conductivity of field lines and therefore the intensity of field-aligned currents seem to be insufficient to provide a current system of the intensity observed. Taking this result into account we will consider the geomagnetic bays to be caused in the main by an ionospheric current system.

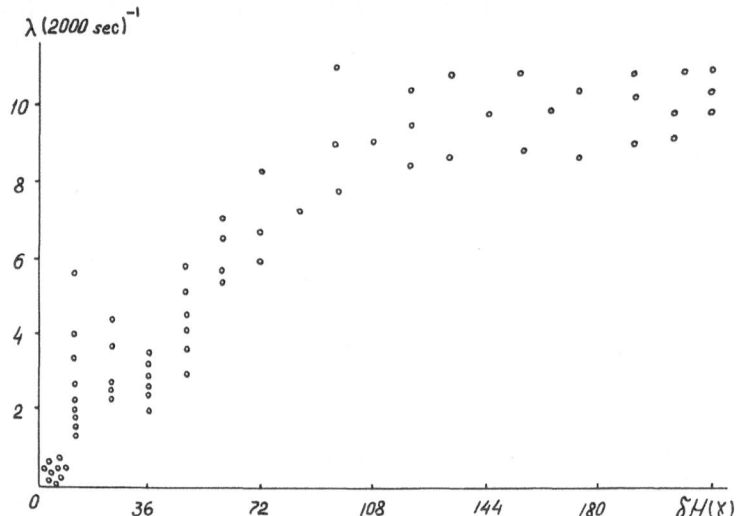

Fig. 3.   The dependence of the decrement of high-latitude geomagnetic pulsations on the intensity of geomagnetic bays (after Pudovkin and Barsukov, 1970).

3.2. ELECTRIC FIELD OF THE DP SYSTEM

3.2.1. *Field in the Auroral Electrojet*

The determination of the field is relatively simple in this region, where good experimental data on the electric currents exist (Meredith *et al.*, 1959). One can see from Table I that the currents in the auroral zone are determined mainly by Hall conductivity. Therefore the electric field in an auroral arc must be directed approximately along the magnetic meridian (equatorward at night and poleward by day).

It must be remembered however that from the analysis of the geomagnetic disturbances one cannot obtain the distribution of the real current density in the ionosphere, but only the disturbance component, that is the difference between the current at the given moment and during quiet conditions. As a result, the electric field calculated is not the total field E but is $\delta E = E - E_0$, where E_0 is the undisturbed field. Thus one may expect some discrepancy between the E field derived from the analysis of the DP current system, and that measured directly. Let us consider, as an example, the Cole (1960) model of the auroral electrojet shown in Figure 4.

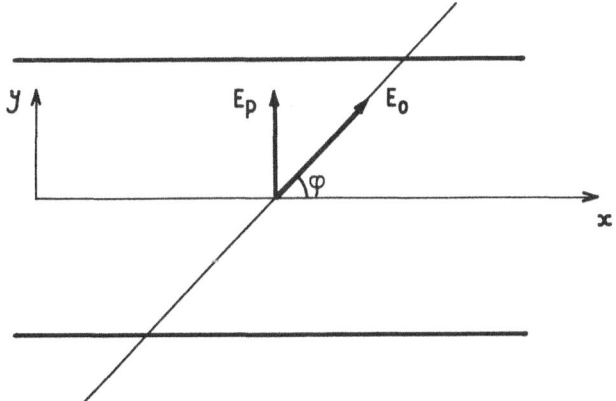

Fig. 4. Fields in the auroral electrojet.

In this figure E_p is the polarization field and φ is the angle between the direction of the current belt and the original field E_0. It follows from the continuity equation div $J = 0$ that inside the belt of enhanced conductivity the polarization field is given by

$$
\begin{aligned}
E_{px} &= 0 \\
E_{py} &= \{(\Sigma'_H - \Sigma_H) \cos \varphi - (\Sigma'_P - \Sigma_P) \sin \varphi\} \, E_0 / \Sigma'_P \\
&\approx E_0 \{(\Sigma'_H / \Sigma'_P) \cos \varphi - \sin \varphi\}
\end{aligned} \tag{6}
$$

and the total field by

$$
\begin{aligned}
E_x &= E_0 \cos \varphi \\
E_y &= E_0 \sin \varphi + E_{py} \\
&\approx (\Sigma'_H / \Sigma'_P) \, E_0 \cos \varphi \, .
\end{aligned} \tag{7}
$$

The approximations hold when the enhanced conductivities Σ'_P and Σ'_H are respectively much greater than the original values Σ_P and Σ_H. In this case the disturbance current density within the electrojet is given by

$$
\begin{aligned}
j_x - j_{x0} &= \{\Sigma'_P + (\Sigma'_H)^2 / \Sigma'_P\} \, E_0 \cos \varphi \\
j_y - j_{y0} &= 0 \, .
\end{aligned} \tag{8}
$$

It is seen from Equations (6)–(8) that the disturbance current is always parallel to the belt boundaries and the polarization field is always perpendicular to the belt (hence to the current too).

Unlike the polarization field, the total field is not necessarily perpendicular to the current sheet, and its direction depends on the relation between the Hall and Pedersen conductivities. With regard to the field obtained by direct measurements (e.g. by means of barium vapour releases), its direction depends on the nature of the original field E_0:

(1) If E_0 is an electrostatic field (it may be a field transmitted from the magnetosphere), the measured field is equal to the total field E and the angle ψ between E and the current jet is determined by

$$\tan \psi = \Sigma'_H/\Sigma'_P.$$
$$\text{For } \Sigma'_H/\Sigma'_P \approx 2 \text{ (see Table I), } \psi \approx 65°. \tag{9}$$

(2) If E_0 is a dynamo-field which is in its nature a local field and cannot be transmitted along the magnetic field line, the measured field must be equal to the polarization field E_p. In this case the measured field within the electrojet is always perpendicular to the disturbance current $j-j_0$, as if the ionospheric currents were predominantly Hall currents, independently of the relative magnitudes of Σ'_P and Σ'_H.

The results mentioned above have been obtained by assuming the enhanced conductivity sheet to be infinite in length. Since in reality this layer is limited in length, the problem is more complicated. Nevertheless if the enhanced conductivity region is sufficiently extended the above conclusions still hold. The corresponding calculations were carried out by many authors: by Fukushima (1953), Weaver and Skinner (1960), Kim and Kim (1964), Asaulenko and Pudovkin (1965). In these papers the region of enhanced conductivity was assumed to be elliptical in shape. Figure 5 shows the intensity of the polarization field (in terms of E_0) in the directions parallel and perpendicular to the major axis of the ellipse, as a function of the Hall conductivity

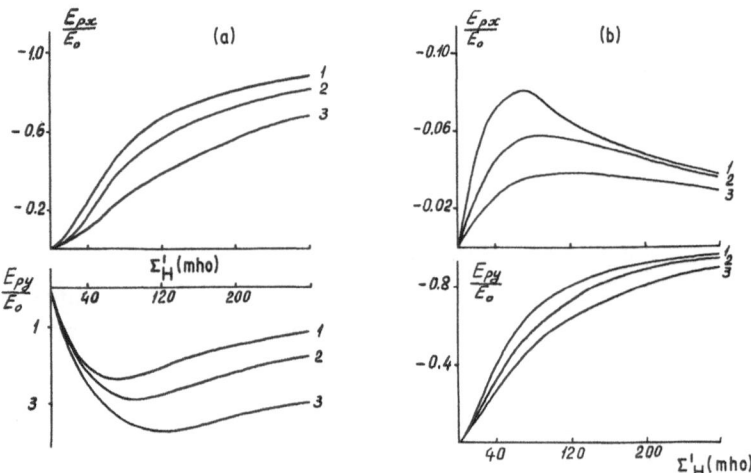

Fig. 5.   Polarization fields in an elliptical region of enhanced conductivity: (a) E_0 parallel to major axis of ellipse, and (b) E_0 perpendicular to major axis of ellipse.
In each diagram, curve 1 refers to an ellipse axis ratio of 30; curve 2 to a ratio of 50, curve 3 to a ratio of 100.

Σ'_H. The original field E_0 is assumed parallel (Figure 5a) or perpendicular (Figure 5b) to this axis. It is seen from the figure that when the Hall conductivity in the ellipse varies in the range of 20–60 mho and the ratio of the ellipse axes is between 30 and 100, the polarization field E_p is approximately perpendicular to the major axis.

The direction of the original field E_0 within the electrojet is difficult to establish, since the disturbance current $(j-j_0)$ always flows along the auroral arc. However it is not difficult to show that in the case when the whole auroral zone is disturbed the current in the electrojet is parallel to the original field E_0 (Fukushima, 1953; Loginov, 1965). Hence the original field E_0 in the electrojet is directed along the lines of magnetic latitude: westward at night and eastward in the evening.

The mean intensity of the E field within the electrojet may be estimated both from the value of the ratio j/Σ'_H (Obayashi and Nishida, 1968) and from the velocity of the electron movement along the current sheet (Akasofu and Chapman, 1961; Boström, 1964). Both methods give similar results: $E \approx 20$ mV/m, the original field E_0 being 2–5 times smaller.

However, the intensity of the electric field is not constant and varies significantly with local time and also from case to case. In order to follow these variations one has to know both the current density and conductivity of the ionosphere at any moment. This may be carried out by means of simultaneous measurements of the magnitude of the geomagnetic disturbance (δH) and the intensity of the cosmic noise absorption (δA) (Pudovkin *et al.*, 1964; Pudovkin, 1968). Thus, the cosmic noise absorption equals

$$\delta A = C_1 N_1 v_1 \, \Delta h_1, \tag{10}$$

where N_1 and v_1 are the average electron density and collision frequency in the absorbing layer, Δh_1 is the thickness of this layer, and C_1 is a constant.

On the other hand, the intensity of the geomagnetic disturbance is equal to:

$$\delta H = C_2 N_2 E \, \Delta h_2 \tag{11}$$

where N_2 is the electron density in the current sheet, Δh_2 is the thickness of this sheet, and C_2 is another constant.

It follows from these equalities that if the electron densities in both layers vary synchronously, the electric field equals

$$E = C_3 v_1 \, \frac{\delta H}{\delta A}. \tag{12}$$

The value of v_1 seems to change insignificantly in the course of a single magnetic bay, and hence variations of the ratio $\delta H/\delta A$ are proportional to the variations of E.

Figure 6 shows an example of the variations of δH, δA and $\delta H/\delta A$ in the course of a magnetic disturbance on 17th March 1963 at Murmansk, according to the data by Pudovkin *et al.* (1964). It is seen that the E field, although rather variable during the disturbance, does not vanish in the intervals between the peaks of δH and does not greatly increase during those peaks. So the variations of δH during a storm seem to be caused by the variations of the conductivity of the ionosphere rather than by the variations of the electric field intensity. The proportionality between the values of δH and δA is also found for many separate events, in which the magnitude of δT_H, the total horizontal disturbance component, varies in the range from 60γ to 700γ (see Figure 7 which shows data given by Pudovkin *et al.*, 1964). This fact suggests that the intensity of the electric field in the ionosphere in the auroral zone is almost independent of the level of geomagnetic activity. This result is hardly comprehensible if the electric field is transmitted into the ionosphere from the magnetosphere, the state of which, including the plasma convection velocity, is clearly dependent on the level of geomagnetic activity (Behannon and Ness, 1966; Heppner *et al.*, 1967). Thus the electric field of the polar storm appears to originate within the ionosphere and is due to some process whose intensity is independent of the level of magnetospheric agitation. As the wind velocity in the dynamo-layer is also independent of geomagnetic activity (Mirkotan, 1962), dynamo-action of the ionospheric winds may be a suitable mechanism (Pudovkin, 1966).

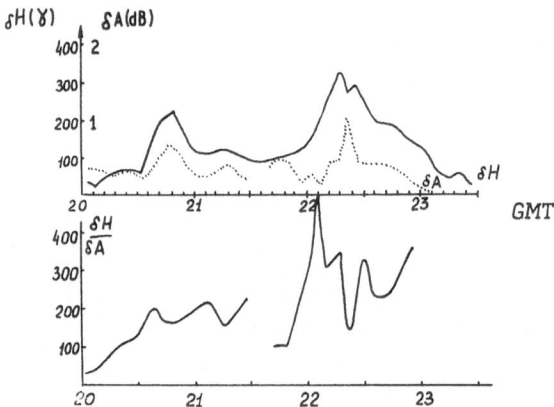

Fig. 6.   Variation of δH, δA and $\delta H/\delta A$ during disturbance of 17th March 1963. (After Pudovkin *et al.*, 1964.)

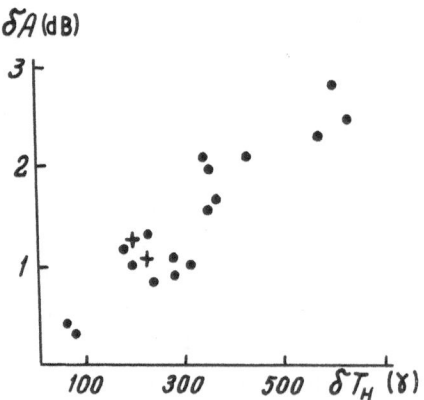

Fig. 7.   Variation of δA with δT_H (after Pudovkin *et al.*, 1964)
• • Negative bays; + + Positive bays.

The diurnal variation of the electric field intensity may be obtained from the diurnal variation of the ratio $\delta H/\delta A$. However, as the energy of the particles precipitated into the higher atmosphere, and consequently the height of the region of the anomalous ionization, have a significant diurnal variation (Barcus and Rosenberg, 1966; Zaitzeva et al., 1969), the value of v_1 in Equation (12) cannot be assumed to be constant, and E is proportional to the product $v_1(\delta H/\delta A)$. The diurnal variation of this quantity at Murmansk is shown according to the data of Pudovkin (1968) in Figure 8. It is seen from the figure that the intensity of the electric field changes sign at 19–20 h LT and reaches its maximum near midnight.

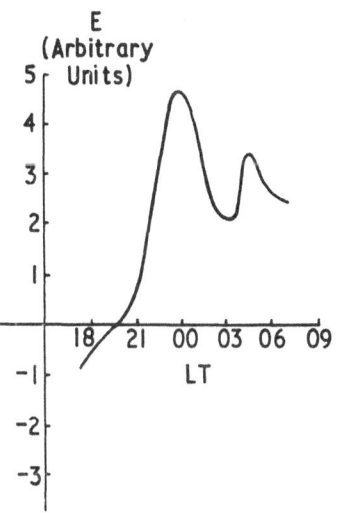

Fig. 8.   Diurnal variation of disturbance electric field at Murmansk (after Pudovkin, 1968).

It is interesting to note that at some other observatories which are also located in the auroral zone, the diurnal variation of E-field intensity may be different. For instance, at College (Alaska) the E-field changes sign at 23–24 h LT and reaches its maximum at 03–04 h LT (Loginov et al., 1962).

3.2.2. Field in the Polar Cap

The picture is much more complicated in the polar cap. The intensity of geomagnetic disturbances in this region is not high (at least in the winter) and does not exceed $100–150\gamma$; hence, the contribution of field-aligned currents to the disturbance field may be relatively high.

We shall first consider the polar cap electric field under the assumption that a geomagnetic bay is completely determined by an ionospheric current system as shown in Figure 2a or 2b. It should first be noted that the current system built up from the magnetic data has some uncertainty, especially near the auroral zone. This uncertainty originates from the fact that near the auroral zone the field measured at the earth's surface consists of two components: the field of the electrojet and the field of the return currents. As the currents flow in the ionosphere at a height of about 100–130 km, the point where the currents change their sign may not coincide with the point where the H-component of the magnetic field reverses. It may be shown that in a typical case with an electrojet of width 200 km, flowing at a height of 100 km, the points where $\delta H = 0$ are 300–400 km distant from the electrojet boundaries. Consequently the focus of the current vortex shown in Figure 2b must actually be located 300–400 km nearer to the electrojet, which makes this current system more like that of Figure 2a.

These systems are alike on the whole, but there is an important difference between them. This is that the foci of the current vortices are located in the first case at a latitude of about 70°, that is adjacent to the auroral zone, while in the second case the focus is located at a latitude of 75°–80° and is not associated with the auroral zone. The physical meaning of this difference may be seen as follows. Writing Ohm's law in the form

$$j = \Sigma_P E + \Sigma_H (B \times E)/B, \tag{13}$$

we obtain in a time-independent case with homogeneous magnetic field

$$\operatorname{curl} j = \nabla \Sigma_P \times E - (B \Sigma_H \operatorname{div} E)/B. \tag{14}$$

This shows that a current system focus (where $\operatorname{curl} j \neq 0$) may occur in either of two significantly different regions:

(1) If the vortex is formed by Hall currents, its focus may be located at any point where an electrostatic charge exists (div $E \neq 0$), and need not be adjacent to the auroral zone;

(2) If the vortex is formed by Pedersen currents, its focus must be located in a region where $\nabla \Sigma_p \neq 0$, that is, most probably, near the boundary of the auroral zone.

Thus one can see that the current systems shown in Figures 2a and 2b may correspond to quite different physical conditions in the ionosphere. Particularly, if the current system 2b occurs, it would probably mean that the auroral substorm currents are mainly Hall currents. In such a case the E field in the polar cap is perpendicular to the current lines and has the configuration proposed by Axford and Hines (1961) and by Taylor and Hones (1965), being directed from the focus of the morning vortex and toward that of the evening vortex, that is along the meridian 03–15h LT near the geomagnetic pole.

However the assumption that the high-latitude ionosphere has a purely Hall conductivity is hardly consistent with present ideas concerning the values of Σ_H and Σ_P (see Table I). The current system of Silsbee and Vestine (1942), as shown in Figure 2a, might fit these ideas much better.

A detailed analysis of the substorm current system, carried out by Akasofu and Meng (1969) shows that the Silsbee and Vestine (1942) current system is the more consistent with the experimental data.

Thus Pedersen current may be an important component of the polar cap current system. As a result the distribution of electric field over the cap must differ significantly from that in the Taylor and Hones (1965) model. However, when calculating

the distribution of the electric field in the polar cap, one must bear in mind that, as noted above, one can obtain only the disturbance component of the current density from the magnetic data.

Two aspects of the calculated field are illustrated in Figure 9. Figure 9a shows the distribution of the disturbed component of the electric field, using the current system of Figure 2a. The electric field of the disturbance is directed from the evening side of the cap to the morning side along the meridian 21–09 h LT and its intensity is about 20 mV/m.

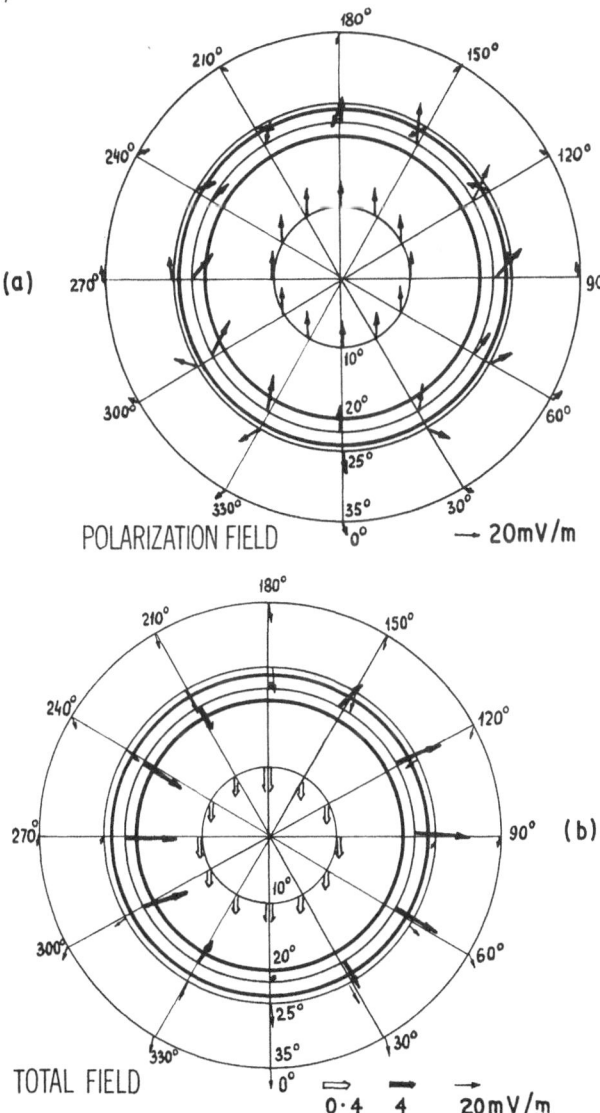

Fig. 9a–b. The disturbance electric field in the polar cap (a) Polarization field (b) Total field. In each case, the azimuth scale represents directions relative to that of E_0. On this scale midnight occurs at 45°.

Figure 9b, obtained on the basis of the same current system, shows the total electric field E, the original electric field E_0 being assumed homogeneous and directed along the meridian 9–21 h LT. It is seen that the total electric field E in the polar cap is directed from the morning side of the earth to the evening side, that is antiparallel to the current of the disturbance. It is parallel to, but much smaller than E_0.

If the original field E_0 is a dynamo-field, the measured field will, as was mentioned above, be equal to the polarization field E_p.

The picture becomes much more complicated when the influence of field-aligned currents is taken into account. However it may be shown that the electric field in the polar cap must have a configuration of the type 9b when the Birkeland current system (Figure 2c) applies, and of the type 9a if the Boström system (Figure 2d) occurs.

3.2.3. *Field in Middle and Low Latitudes*

The calculation of the polar storm electric field in middle and low latitudes is of a rather formal character as it is not clear if the middle-latitude currents of a storm really exist. However as the data mentioned in Section 3.1 may be considered as evidence that such return currents do exist, the calculation is of some interest. Some ideas on the configuration of the DP polarization electric field may be obtained from Figure 9a. It is seen that the intensity of the E_p field has a maximum near the auroral zone and rapidly decreases with distance from this zone.

Using a more realistic model of the ionosphere, calculations of the E_p field of a polar storm were carried out by Maeda (1959). The results obtained in this paper are shown in Figure 10. It is seen that inside the auroral zone E_p is as high as 20 mV/m, but this rapidly decreases at latitudes below 70° to 5 mV/m at the geomagnetic equator.

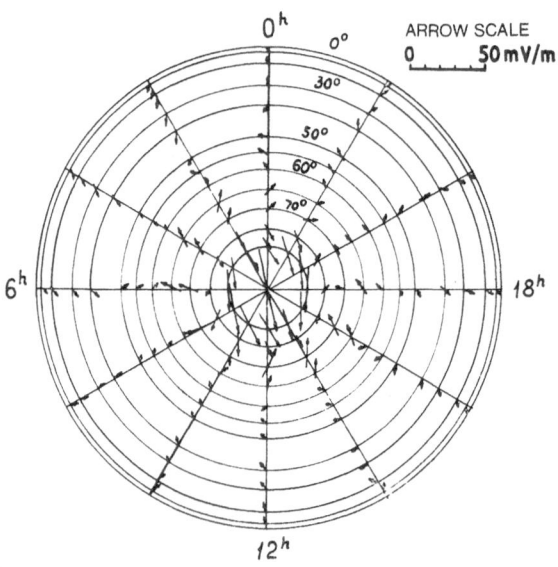

Fig. 10.   The disturbance polarization field, viewed from above the N pole (After Maeda, 1959).

4. Electric Field Effects on Ionization Distributions

4.1. LARGE-SCALE DISTRIBUTIONS

The most important effect of electric fields in influencing the large-scale distribution of ionization is that arising from electrodynamic drift in the presence of the earth's magnetic field. This drift contributes to plasma movement which may constitute an important term in the continuity equation that determines the electron distribution. In general this equation may be written

$$\frac{\partial N}{\partial t} = Q - L - \text{div}(Nv) \tag{15}$$

where N is the electron number density, Q and L are respectively the production and loss rates, and v represents the drift velocity of either ions or electrons. This drift velocity is determined by the equation of motion of the relevant species. Considering the ions for instance, of mass m_i and charge e, temperature T and effective collision frequency v with neutral particles, this equation is

$$\frac{dv}{dt} = g + \frac{e}{m}(E + v \times B) \frac{k}{Nm_i} \nabla(NT) + v(u - v) \tag{16}$$

where k is Boltzmann's constant, g is the gravitational acceleration and u is the velocity of the neutral gas.

The ion velocity required for substitution into Equation (15) thus depends on the motion of the neutral gas, as well as on the unknown electron density. In order to obtain N, the simplest procedure is to assume some wind system, i.e. to regard u as being prescribed, and then to solve Equations (15) and (16) simultaneously for N and v. However since the wind system is not well known, it is preferable to consider also the equation of motion of the neutral gas,

$$\frac{du}{dt} = g - \frac{1}{nm} \nabla p + \frac{Nv}{n}(v - u) + 2u \times \omega + \frac{\mu}{nm} \nabla^2 u \tag{17}$$

where n and m are the neutral particle number density and mass respectively, p is the gas pressure and μ is the coefficient of viscosity. The procedure is then to obtain simultaneous solutions of Equations (15), (16) and (17), giving self-consistent values of electron density and wind and plasma drift speeds.

4.1.1. F Region

A large number of theoretical investigations have been carried out in recent years on the effects of electric fields in the F region. These have all been based on solutions of Equations (15) and (16), but with varying degrees of sophistication as to the inclusion of time-dependence and the motion of the neutral gas. The main attention has been directed towards the equatorial zone, and in particular to explaining the morphology of the equatorial anomaly. It was soon realized that the use of Equations (15) and (16), with the electric field neglected, would not yield a sufficiently pro-

nounced equatorial anomaly (Rishbeth *et al.*, 1963), but by following the suggestion
of Martyn (1955) and further work by Duncan (1960), it was found that the inclusion
of an eastward electric field of a few hundred microvolts per metre could give quite
realistic daytime ionization distributions in the meridian plane (Bramley and Peart,
1965). The immediate effect of this electric field, which is consistent with the pre-
dictions of the atmospheric dynamo theory as described in Section 2.3, is to produce
an upward plasma drift. Figure 11 shows the very large effect on the ionization
distribution of an eastward electric field of about 800 μV/m, which gives an upward
drift of about 20 m/s.

These early calculations were based on steady-state solutions of Equations (15)
and (16), in which, moreover, air motion was not considered, i.e. u was put equal to
zero. More recent calculations (Hanson and Moffett, 1966; Bramley and Young, 1968;
Varnasavang, 1967; Baxter and Kendall, 1968; Kendall and Windle, 1968; Abur-Robb
and Windle, 1969; Abur-Robb, 1969; Sterling *et al.*, 1969) have removed these re-
strictions in various degrees, though none has yet been carried out for the equatorial
region using all three Equations (15)–(17) as described above. All these investigations
have confirmed that east–west electric fields of the magnitudes mentioned play a
vital part in determining the equatorial F-region electron distribution. It appears
that with a proper combination of winds and fields most of the features of the spatial
and temporal ionization distribution can be reproduced. In the immediate vicinity
of the equator the field is the most important parameter; in particular the detailed
form of the diurnal variations of electron density is critically dependent on the phase
of the field, which reverses direction at night. Indeed the degree of agreement between
experimental data and theoretical results using dynamo fields may be regarded more
as a measure of the accuracy of the dynamo calculations than that of the F-region
theory. We may note that incoherent scatter measurements are now capable of giving
independent data on plasma drift speeds in the F region, from observations of Doppler
frequency shifts. Such measurements at the equator have shown very good correlation
of the F-region vertical drift speed with the electric field in the E region below,
as indicated by the electron drift speed in the equatorial electrojet. They have also
indicated some significant departures of the diurnal variation of the electric field from
that predicted by dynamo theory (Balsley and Woodman, 1970). The correlation of
the equatorial anomaly parameters with the electric field driving the electrojet has
also been demonstrated in studies which use magnetogram data as a monitor of the
jet current (Dunford, 1967, 1970; MacDougal, 1969).

Electric field effects on the electron distribution in the middle-latitude F region
have not received as much attention as in the equatorial case. This is because the effects
depend essentially on vertical ionization drift, which is proportional to the cosine
of the angle of dip and inversely proportional to the magnetic induction. Hence the
effect of an electric field of given magnitude may be expected to diminish with
increasing latitude. Figure 12 shows some theoretical results (Bramley, 1969) for
the diurnal variation of the critical frequency foF2 and height of the peak electron
density h_m calculated for a latitude of 45° using the east–west electric field calculated

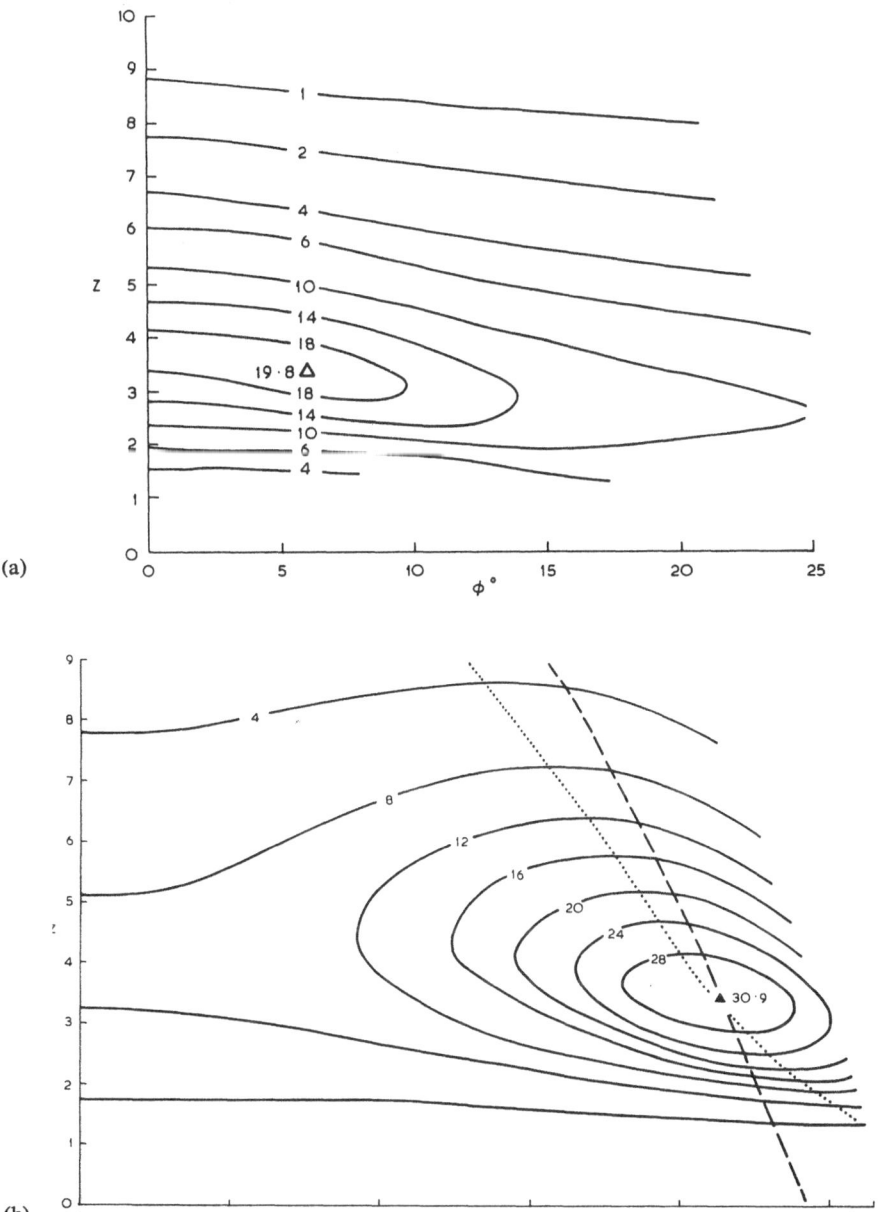

Fig. 11a, b. Electron density distribution as a function of magnetic latitude φ and height z. Unit of electron density $= 10^{11}\,\mathrm{m^{-3}}$; unit of $z = 82$ km. (a) Electric field $= 0$ (after Rishbeth *et al.*, 1963). (b) Electric field $\approx 800\ \mu\mathrm{V/m}$ (after Bramley and Peart, 1965) $\cdots\cdots$ Locus of maximum electron density at different heights --- Geomagnetic fieldline.

by Maeda (1963) from dynamo theory, and including also a typical wind system (Geisler, 1967). These results show that the electric field effects are quite small by day, but somewhat larger by night. They are much smaller than those produced by the wind.

There is another mechanism which may be expected to reduce the electric field effects still further. This is the ion drag mechanism (Hirono and Kitamura, 1956; Dougherty, 1961); the action of the moving ions on the neutral air motion, by collisional interaction, has the theoretical result that in the steady state the vertical

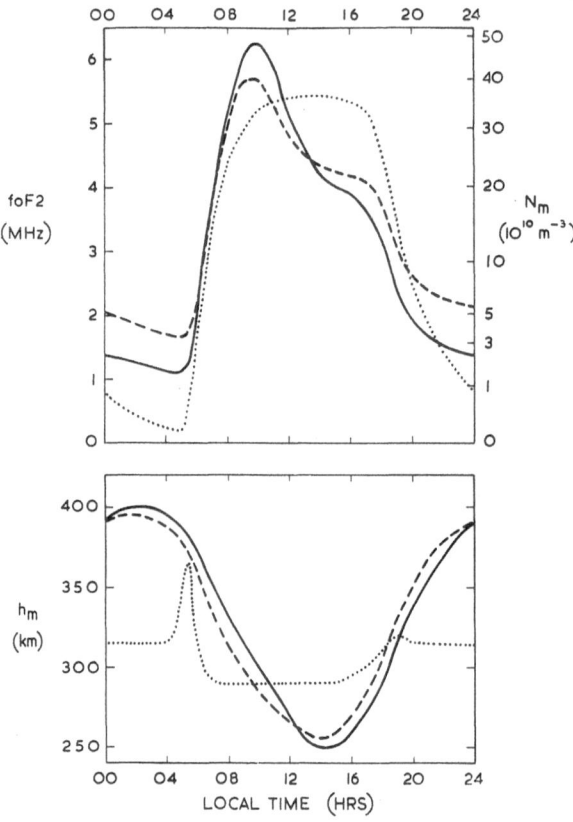

Fig. 12. Diurnal variation of foF2 and h_m at latitude 45° (After Bramley, 1969); ····· No wind or field included; — — — — Including wind but not field; ——————— Including wind and field.

drift due to the electric field is completely annihilated. Thus it is important to determine to what extent this mechanism is effective in a practical time-varying situation. Simple analytical investigations (Rishbeth *et al.*, 1965; Bramley, 1969) suggest that the reduction of vertical drift should be quite effective in the daytime, but much less so at night, as a result of the much lower electron density. Detailed numerical calculations (Bramley and Rüster, in preparation) based on solution of the three coupled Equations (15)–(17), have confirmed that the field effects are not greatly reduced at night. The reduction is certainly greater by day, but in the particular case considered it was not so large as indicated by the simple theory.

It should be noted that the night-time electric field predicted by dynamo theory is in such a direction as to give a downward drift of ionization, and hence a reduction in the peak electron density. Such an electric field can, therefore, only accentuate the problem of the night-time maintenance of the F layer. Stubbe (1968) showed that an electric field of order 5 mV/m could account for this maintenance, but the field would have to be directed towards the east, whereas all the dynamo calculations indicate a westward field at night. Stubbe also showed that a southward field would help in arresting the decay of the F layer by transporting ionization eastward from the day side to the night side, and a field of 10 mV/m would have an appreciable effect. This would only apply to times at which there is a large east–west ionization gradient. Extremely large north–south fields, of order 100 mV/m, have been recorded in the auroral region, but no detailed calculations of the effects of such fields have been published as yet.

In addition to the modifications in the ionization distribution which can be caused by electric fields, significant changes in the neutral air wind system may be produced at the same time, as a result of the transfer, by collisions, of ion motion to the neutral particles. It has been shown by Bramley (1967) that typical dynamo fields may be expected to give rise to both zonal and meridional winds of up to about 40 m/s. The large auroral fields mentioned above should produce much larger wind speeds if they are maintained for a few hours.

4.1.2. E Region

Electromagnetic drift effects in the E region, due to the electric fields which drive the various current systems at these levels, are smaller than the F-region drifts discussed above, because of the much greater density of the neutral atmosphere. Nevertheless, such drift effects have been cited to account for certain perturbations in critical frequency, and also possibly in the height, of the E layer (Beynon and Brown, 1959).

4.2. EFFECTS ON IRREGULAR STRUCTURES

The effect of electric fields on ionization irregularities is of considerable importance, since observations of the movements of such irregularities are frequently made, both by radio and by optical methods, and used to make deductions about neutral air winds and electric fields in the region concerned. Theoretical investigations of the motion of irregularities have been carried out by a number of authors (Martyn, 1953; Kato, 1963, 1964, 1965; Clemmow et al., 1955; Tsedilina, 1965; Haerendel et al., 1967; Kaiser et al., 1969; Kato and Sakurai, 1970). Most attention has been paid to cylindrical irregularities with axis along the geomagnetic field direction, and it has been shown that in idealized conditions the irregularity can move at right angles to the field without change of form. At heights above the E region the drift velocity of a very weak irregularity is given to a first approximation by $E \times B/B^2$, irrespective of the wind speed. However Haerendel et al. (1967) have shown that the use of densely ionized artificial ion clouds may introduce complications by significantly affecting the height-integrated Pedersen conductivity for a field line passing through the cloud. The drift speed then depends on the neutral wind speed, and although this can be measured simultaneously by observation of a neutral luminous cloud produced along with the ionized cloud, an appreciable uncertainty in interpretation can arise from lack of knowledge of the integrated conductivity. Another effect is that significant polarization of the cloud can be produced, giving rise to distortions which have been observed in practice (Haerendel *et al.*, 1969).

Kaiser *et al.* (1969) considered particularly the movement of plane and cylindrical irregularities, inclined arbitrarily to the magnetic field, at E region heights. They showed that the drift speed depends strongly on the inclination to the field; closely field-aligned irregularities move with the electron Hall drift speed, i.e. $E \times B/B^2$, while non-aligned irregularities tend to move with the much smaller Hall drift speed of the ions. The east-west movement of the closely field-aligned irregularities which produce the radio aurora may thus be interpreted as a Hall electron drift.

4.3. IRREGULARITY-PRODUCING EFFECTS

The ionosphere contains a large number of types of spatial irregularity in the electron density, and many different theories have been put forward to account for them. Here we shall discuss briefly two separate types of irregularity-producing mechanism in which electric fields play an important part.

The first case occurs in the equatorial electrojet, where field-aligned irregularities of very small scale, of the order of metres, have been detected by the scattering of h.f. and v.h.f. radio signals. These irregularities are correlated with the strength of the electrojet (Cohen and Bowles, 1963), and take the form of an angular spectrum of plane waves moving perpendicularly to the magnetic field. It appears to be fairly well established that they arise as the result of a two-stream instability mechanism (Farley, 1963), which operates when the relative velocity of ions and electrons exceeds a critical value which is approximately the thermal speed of the ions. It thus occurs when the east-west electric field which drives the electrojet exceeds a certain value, which is of the order of a few hundred microvolts per metre.

The second irregularity-producing mechanism is that which involves crossed electric and magnetic fields in the presence of a gradient in the background ionization density. It was originally studied by Simon (1963), and its ionospheric applications have subsequently been investigated by a number of authors (Tsuda *et al.*, 1966; Sato *et al.*, 1968; Reid, 1968; Tsuda *et al.*, 1969). The instability depends essentially on a difference between the ion and electron mobilities, and depending on the geometry of the fields and the density gradient it may be either the Hall or the Pedersen mobility which is involved. In either case, irregular polarization fields are developed, which interact with the magnetic field to produce drifts in such a sense as to amplify the irregularities. The main application of this theory has been to explain small-scale irregularities in the E region, but Reid (1968) has pointed out the relevance of the mechanism to F-layer irregularities also, in particular the small-scale field-aligned structures which produce spread-F reflections. It is suggested that these structures are caused by the transfer of irregular electric fields into the F region along the highly conducting field lines, from the E region where the irregularities are most efficiently produced by the cross-field mechanism. This theory is thus closely related to the earlier suggestion of Dagg (1957) that F-region irregularities could similarly arise from small-scale electric fields generated in the E region by turbulent wind motion. However Reid's theory avoids some difficulties which are inherent in this mechanism.

A similar cross-field instability was proposed for F-layer irregularities by Martyn (1959), who showed that irregularities on the underside of the layer would be amplified at a time when the layer is undergoing upward electrodynamic drift, because the polarization charges set up on the boundary of the irregularity would cause it to move at a slightly different speed from that of the main layer of ionization. The validity of this theory is questionable (Dougherty, 1959; Fejer, 1959) on the grounds that the polarization fields would tend to be short-circuited by the E region, so that no differential drift speed would exist. A similar mechanism has recently been invoked however (Haerendel *et al.*, 1969) to explain the production of striations in artificial ion clouds. As in Martyn's case, the theory indicates that these would develop only at the trailing edge of the cloud, as has been found experimentally. The question of possible short-circuiting of the polarization charges by the E region again arises, however, and Reid (1970) has suggested that the observed striations result from irregular fields transmitted from the E region below rather than being generated at the cloud itself.

Acknowledgements

The paper is published with the permission of the Director of the Radio and Space Research Station of the Science Research Council.

References

Abur-Robb, M. F. K.: 1969, *Planetary Space Sci.* **17**, 1269.
Abur-Robb, M. F. K. and Windle, D. W.: 1969, *Planetary Space Sci.* **17**, 97.
Afonina, R. G. and Feldstein, Y. I.: 1970, *Aurorae* **19**, 77, Nauka, Moscow.
Akasofu, S. I.: 1966, *Space Sci. Rev.* **6**, 21.
Akasofu, S. I. and Chapman, S.: 1961, Sci. Rept. No. 7, National Science Foundation Grant No. Y/22 6/327.
Akasofu, S. I. and Meng, C. I.: 1969, *J. Geophys. Res.* **74**, 293.
Asaulenko, L. G. and Pudovkin, M. I.: 1965, *Geomagnetizm i Aeronomiya* **5**, 247.
Axford, W. I. and Hines, C. O.: 1961, *Can. J. Phys.* **39**, 1433.
Baker, W. G. and Martyn, D. F.: 1953, *Phil. Trans. Roy. Soc.* A **246**, 281.
Balsley, B. B. and Woodman, R. F.: 1970, in press.
Barcus, J. R. and Rosenberg, T. J. 1966, *J. Geophys. Res.* **71**, 803.
Baxter, R. G. and Kendall, P. C.: 1968, *Proc. Roy. Soc.* A **304**, 171.
Behannon, K. W. and Ness, N. F.: 1966, *J. Geophys. Res.* **71**, 2321.
Beynon, W. J. G. and Brown, G. M.: 1959, *J. Atmospheric Terrest. Phys.* **14**, 138.
Birkeland, K.: 1908, *The Norwegian Aurora Polaris Expedition*, 1902–1903, Christiania.
Boström, R.: 1964, *J. Geophys. Res.* **69**, 4983.
Boström, R.: 1968, *Ann. Geophys.* **24**, 681.
Bramley, E. N.: 1967, *J. Atmospheric Terrest. Phys.* **29**, 1317.

Bramley, E. N.: 1969, *J. Atmospheric Terrest. Phys.* **31**, 1223.
Bramley, E. N. and Peart, M.: 1965, *J. Atmospheric Terrest. Phys.* **27**, 1201.
Bramley, E. N. and Young, M.: 1968, *J. Atmospheric Terrest. Phys.* **30**, 99.
Brünelli, B. E.: 1963, *Geomagnetizm i Aeronomiya* **3**, 746.
Cahill, L. J.: 1959, *J. Geophys. Res.* **64**, 1377.
Chamberlain, J. W.: 1961, *Astrophys. J.* **134**, 401.
Chapman, S.: 1956, *Nuovo Cimento* **4**, 1385.
Clemmow, P. C., Johnson, M. A., and Weekes, K.: 1955, in *Physics of the Ionosphere*, Physical Society, London, p. 136.
Cohen, R. and Bowles, K. L.: 1963, *J. Geophys. Res.* **68**, 2503.
Cole, K. D.: 1960, *Australian J. Phys.* **13**, 484.
Dagg, M.: 1957, *J. Atmospheric Terrest. Phys.* **11**, 139.
Dougherty, J. P.: 1959, *J. Geophys. Res.* **64**, 2215.
Dougherty, J. P.: 1961, *J. Atmospheric Terrest. Phys.* **20**, 167.
Duncan, R. A.: 1960, *J. Atmospheric Terrest. Phys.* **18**, 89.
Dunford, E.: 1967, *J. Atmospheric Terrest. Phys.* **29**, 1489.
Dunford, E.: 1970, *J. Atmospheric Terrest. Phys.* **32**, 421.
Dungey, J. W.: 1963, in *Geophysics: The Earth's Environment*, Gordon and Breach, p. 505.
Farley, D. T.: 1963, *J. Geophys. Res.* **68**, 6083.
Fatkullin, M. N.: 1964, *Geomagnetizm i Aeronomiya* **4**, 464.
Fejer, J. A.: 1953, *J. Atmospheric Terrest. Phys.* **4**, 184.
Fejer, J. A.: 1959, *J. Geophys. Res.* **64**, 2217.
Föppl, H., Haerendel, G., Haser, L., Lüst, R., Melzner, F., Meyer, B., Neuss, H., Rabben, H. H. Rieger, E., Stöcker, J., and Stoffregen, W.: 1968, *J. Geophys. Res.* **73**, 21.
Fukushima, N.: 1953, *J. Fac. Sci. Univ. Tokyo* **11**, 8.
Geisler, J. E.: 1967, *J. Atmospheric Terrest. Phys.* **29**, 1469.
Haerendel, G., Lüst, R., and Rieger, E.: 1967, *Planetary Space Sci.* **15**, 1.
Haerendel, G., Lüst, R., Rieger, E., and Völk, H.: 1969, in *Atmospheric Emissions*, Van Nostrand Reinhold, p. 293.
Hanson, W. B. and Moffett, R. J.: 1966, *J. Geophys. Res.* **71**, 5559.
Heppner, J. P., Sugiura, M., Skillman, T. L., Ledley, B. G., and Campbell, M.: 1967, *J. Geophys. Res.* **72**, 5417.
Hirono, M. and Kitamura, T.: 1956, *J. Geomag. Geoelect.* **8**, 9.
Hirono, M., Maeda, H., and Kato, S.: 1959, *J. Atmospheric Terrest. Phys.* **15**, 146.
Kaiser, T. R., Pickering, W. M., and Watkins, C. D.: 1969, *Planetary Space Sci.* **17**, 519.
Kato, S.: 1956, *J. Geomag. Geoelect.* **8**, 24.
Kato, S.: 1957, *J. Geomag. Geoelect.* **9**, 107.
Kato, S.: 1963, *Planetary Space Sci.* **11**, 823.
Kato, S.: 1964, *Planetary Space Sci.* **12**, 1.
Kato, S.: 1965, *Space Sci. Rev.* **4**, 223.
Kato, S. and Sakurai, H.: 1970, *J. Atmospheric Terrest. Phys.* **32**, 1117.
Kendall, P. C. and Windle, D. W.: 1968, *Geophys. J. Roy. Astron. Soc.* **15**, 147.
Kim, J. S. and Kim, H. Y.: 1962, *Nature* **196**, 630.
Kim, J. S. and Kim, H. Y.: 1963, *J. Atmospheric Terrest. Phys.* **25**, 481.
Kim, J. S. and Kim, H. Y.: 1964, *Can. J. Phys.* **42**, 569.
Kim, J. S. and Volkman, R. A.: 1963, *J. Geophys. Res.* **68**, 3187.
Korablev, A. V. and Rudakov, L. I.: 1966, *J. Experim. Tech. Phys.* **50**, 220.
Langel, R. A. and Cain, J. C.: 1968, *Ann. Geophys.* **24**, 857.
Loginov, G. A.: 1965, *Geomagnetizm i Aeronomiya* **5**, 251.
Loginov, G. A., Pudovkin, M. I., and Skrynnikov, R. G.: 1962, *Geomagnetizm i Aeronomiya* **2**, 709.
MacDougall, J. W.: 1969, *Radio Sci.* **4**, 805.
Maeda, H.: 1957, *J. Geomag. Geoelect.* **9**, 86.
Maeda, H.: 1959, *J. Geomag. Geoelect.* **10**, 66.
Maeda, H.: 1963, in *Proc. Int. Conf. Ionosphere*, Physical Society, London, p. 187.
Maeda, H. and Matsumoto, H.: 1962, *Rep. Ionosph. Space Res.* **16**, 1.
Mange, P.: 1960, *J. Geophys. Res.* **65**, 3833.
Martyn, D. F.: 1953, *Phil. Trans. Roy. Soc. A* **246**, 306.

Martyn, D. F.: 1955, in *Physics of the Ionosphere*, Physical Society, London, p. 254.
Martyn, D. F.: 1959, *Proc. Inst. Radio Engrs.* **47**, 147.
Matsushita, S.: 1953, *J. Geomag. Geoelect.* **5**, 109.
Matsushita, S.: 1969, *Radio Sci.* **4**, 771.
Mayr, H. G. and Volland, H.: 1968, *J. Geophys. Res.* **73**, 4851.
McNish, A. G.: 1937, *Terr. Mag.* **43**, 1.
Meredith, L. H., Davis, L. R., Heppner, J. P., and Berg, O. E.: 1959, in *Experimental Results of the U.S. Rocket Programme for IGY*, Guttmacher, p. 169.
Mirkotan, S. F.: 1962, *Geomagnetizm i Aeronomiya* **2**, 578.
Obayashi, T. and Nishida, A.: 1968, *Space Sci. Rev.* **8**, 3.
Ossakow, S. L.: 1968, *J. Geophys. Res.* **73**, 6366.
Piddington, J. H.: 1960, *J. Geophys. Res.* **65**, 93.
Piddington, J. H.: 1964, *Space Sci. Rev.* **3**, 724.
Pudovkin, M. I.: 1960, *Izvestia Akad. Nauk S.S.S.R., Ser. Geophys.* **3**, 484.
Pudovkin, M. I.: 1966, in *Highlatitude Investigations in Geomagnetism and Aeronomy*, Moscow, p. 19.
Pudovkin, M. I.: 1968, in *Physics of the Magnetosphere and Auroral Substorms*, Irkutsk, p. 99.
Pudovkin, M. I., Skrynnikov, O. I., and Shumilov, O. I.: 1964, *Geomagnetizm i Aeronomiya* **4**, 848.
Pudovkin, M. I. and Barsukov, V. M.: 1970, *Geomagnetizm i Aeronomiya*, in press.
Reid, G. C.: 1965, *J. Res. Nat. Bureau Stds.* **69D**, 827.
Reid, G. C.: 1968, *J. Geophys. Res.* **73**, 1621.
Reid, G. C.: 1970, *Planetary Space Sci.* **18**, 1105.
Rishbeth, H., Lyon, A. J., and Peart, M.: 1963, *J. Geophys. Res.* **68**, 2559.
Rishbeth, H., Megill, L. R., and Cahn, J. H.: 1965, *Ann. Geophys.* **21**, 235.
Sato, T., Tsuda, T., and Maeda, K.: 1968, *Radio Sci.* **3**, 529.
Silsbee, H. C. and Vestine, E. H.: 1942, *Terr. Mag.* **47**, 195.
Simon, A.: 1963, *Phys. Fluids* **6**, 382.
Spreiter, J. R. and Briggs, B. R.: 1961a, *J. Geophys. Res.* **66**, 1731.
Spreiter, J. R. and Briggs, B. R.: 1961b, *J. Geophys. Res.* **66**, 2345.
Sterling, D. L., Hanson, W. B., Moffett, R. J., and Baxter, R. G.: 1969, *Radio Sci.* **4**, 1005.
Stewart, B.: 1882, in *Encyclopaedia Britannica*, 9th ed., London and New York, **16**, 159.
Stubbe, P.: 1968, *J. Atmospheric Terrest. Phys.* **30**, 243.
Swift, D. L.: 1965, *J. Geophys. Res.* **70**, 3061.
Taylor, H. E. and Hones, E. W.: 1965, *J. Geophys. Res.* **70**, 3605.
Tsedilina, Y. Y.: 1965, *Geomagnetizm i Aeronomiya* **5**, 525.
Tsuda, T., Sato, T., and Maeda, K.: 1966, *Radio Sci.* **1**, 212.
Tsuda, T., Sato, T., and Matsushita, S.: 1969, *J. Geophys. Res.* **74**, 2923.
Vanyan, L. L.: 1968, in *Physics of the Magnetosphere and Auroral Substorms*, Irkutsk, p. 230.
Varnasavang, V.: 1967, *J. Geophys. Res.* **72**, 1555.
Weaver, J. T. and Skinner, R.: 1960, *Can. J. Phys.* **38**, 1104.
Zaitzeva, S. A., Ivliev, D. Y., and Pudovkin, M. I.: 1969, *Geomagnetizm i Aeronomiya* **9**, 86.
Zmuda, A. J., Martin, J. M., and Heuring, F. T.: 1966, *J. Geophys. Res.* **71**, 5033.
Zmuda, A. J., Heuring, F. T., and Martin, J. M.: 1967, *J. Geophys. Res.* **72**, 1115.

THE STRUCTURE OF THE PLASMASPHERE ON THE BASIS OF DIRECT MEASUREMENTS

K. I. GRINGAUZ

Radiotechnical Institute of the Academy of Sciences, Moscow, U.S.S.R.

Abstract. The plasmasphere is the outermost region of the ionosphere which extends up to several earth radii. It is filled with a collisionless cold plasma in which the geomagnetic field is frozen. The data obtained from the first measurements in this region (up to 1963) are given as a background. The information from experiments carried out after 1963 by means of charged particle traps, ion mass spectrometers, and Langmuir probes installed on satellites of the Electron, IMP, and OGO series is reviewed. It comprises the space distribution of ion and electron densities, their dependence on geomagnetic activity and local time, ion composition of the plasmasphere, and charged-particle temperatures. The possibility of detection of the plasmapause by the observation of more energetic charged particles is mentioned, and the results of in situ measurements are briefly compared with whistler results.

The importance of plasmasphere study for the geophysics is outlined.

1. Introduction

1.1. Definitions

Generally accepted terminology relating to all parts of the ionosphere, the envelope of ionized gas that surrounds the earth, does not as yet exist. In this review, only the outermost region of the ionosphere is considered; the region filled with collisionless plasma (consisting basically of ionized hydrogen particles – protons and electrons – having thermal energies) with a frozen-in magnetic field. The upper boundary of this region is, on the average, situated at a geocentric distance R of about 4.5 times R_E (the earth's radius). The degree of ionization in this region is very high – at $R \sim 2.5\ R_E$ (height $Z \sim 10000$ km), the density of neutral hydrogen $n_0 = 10^3$ cm^{-3} (Kurt, 1967), while for this case the average ion density n_i is also $\sim 10^3$ cm^{-3} (Bezrukikh and Gringauz, 1965). The term plasmasphere (introduced by Carpenter, 1966) shall be used to denote this specific region in this review. It has often been called the 'protonosphere' (Geisler and Bowhill, 1965; Gliddon, 1966; Nagy *et al.*, 1968), but we shall not use this term, since, for example, the region close to the maximum of the F layer is not called the 'oxygensphere'.

The upper boundary of the plasmasphere will be called the *plasmapause* (Carpenter, 1966).

Observational results of the plasmasphere published prior to 1967, and some of the problems relating to its formation and existence, were briefly summarized at the Washington symposium of the physics of the magnetosphere (Gringauz, 1969; Axford, 1969; Helliwell, 1969).

In the present review a part of the data reported in Washington is used, and the new results, which become recently available, are included.

1.2. Methods of investigating the plasmasphere

Direct observational studies of the plasmasphere are made with the help of such

Dyer (ed.), Solar-Terrestrial Physics/1970: Part IV, 142–164. All Rights Reserved.
Copyright © 1972 by D. Reidel Publishing Company.

instruments as charged-particle traps, ion mass-spectrometers, and Langmuir probes, carried aboard rockets and satellites (Gringauz *et al.*, 1960a; Gringauz *et al.*, 1960b; Gringauz, 1961; Bezrukikh and Gringauz, 1965; Taylor *et al.*, 1965; Serbu and Maier, 1966; Binsack, 1967; Vasyliunas, 1968a; Brinton *et al.*, 1968; Taylor *et al.*, 1968a, 1969; Bezrukikh, 1968; Chappell *et al.*, 1969; Chappell *et al.*, 1970; Harris *et al.*, 1970; Freeman *et al.*, 1970).

All methods of obtaining information on the plasmasphere which use electromagnetic radiation of natural or man-made origin, be it radiation with very low frequency, for example, whistlers, or with very high frequency (incoherent backscatter radar technique), and independent of whether the observations are carried out on earth or on a satellite, are the subject of a review article, elsewhere in this volume (Helliwell, 1972).

Studies of whistlers from ground-based devices (Carpenter, 1963; Carpenter, 1966; and many other papers) made great contributions to the formulation of contemporary ideas on the physics and dynamics of the plasmasphere. The special value of this method lies in the fact that it is possible to make continuous measurements, thus obtaining, comparatively cheaply, vast quantities of initial observational results. However the possibilities of this method are limited; it gives information on the electron densities only in the geomagnetic equatorial plane, and does not give any information on the electron energy.

Information on the details of electron-density distributions outside of the equatorial plane may be obtained from low-frequency electromagnetic radiation data, obtained from satellites (Carpenter *et al.*, 1969).

Satellite studies of so-called ion whistlers give more information about the surrounding plasma (including ion temperature), but this method can be used only at comparatively low heights (Gurnett and Brice, 1966).

Diverse and valuable information on the variation of the ionospheric plasma parameters with height above the point of observation may be obtained by the method of incoherent backscattering of radio waves. The present state of technology, however, does not allow us to investigate the region of heights corresponding to every layer of the plasmasphere, including its most interesting region, the plasmapause.

The limitations, which are inherent in indirect methods of study of the plasmasphere, are mentioned here only so that we may correctly evaluate the place and value of the direct methods of observation, listed above.

With the help of the direct methods, it is possible to define all parameters of the plasmasphere (ion and electron densities, ion mass composition, energy distribution, etc.) at any latitude, and to obtain data for the construction of a full, three-dimensional model of the plasmasphere.

Such measurements, however, in which it is necessary to register and analyze charged particles with the smallest energies found in space, and with rather low densities, entail significant experimental problems. Despite the significant efforts made in pre-flight calibration of the instrument (for example, Brinton *et al.*, 1968), it has not been possible to fully eliminate the influence of all factors that distort the results of direct measurements. One of these, particularly, is the change in electrical potential of the spacecraft during the flight: this influences the spatial and energy distribution of the charged particles in the neighbourhood of the spacecraft in a highly rarefied medium. Photoemission of electrons from the surface of the spacecraft makes the measurements of the electron component of the plasma especially difficult.

In some cases, authors of the experiments quote a low accuracy for the measurements (for example, to a factor of 5 – Taylor *et al.*, 1965; Gringauz, 1969); in other cases, authors prefer not to give the magnitude of the charged-particle density, and restrict themselves to defining the position of the plasmapause (Binsack, 1967).

Although many of the results obtained appear to be indisputable, even when the authors of the direct measurements do not consider it necessary to emphasize the fact that their measurements are rather rough, some caution is necessary when using data from direct measurements.

1.3. First observations on the plasmasphere

The first information on the existence of the cold plasma envelope around the earth, with boundaries at geocentric distances $R \sim 4 \, R_E$, was published in 1960, in the description of the experiments carried out in 1959 on the first Soviet lunar rockets. The simultaneous use of four traps with different potentials retarding positive ions showed that the plasma temperature does not exceed few tens of thousands of degrees (Gringauz *et al.*, 1960a, b).

In April, 1961, at the second COSPAR symposium a report on the structure of the earth's ionized gas envelope was presented. In this report the problem of the upper boundary of the ionosphere was discussed: "Experiments on board the artificial satellites and space probes compel now to seek paths for creating a new theory which would adequately explain the facts we have acquired of late. Among these facts is the significant increase of negative gradients of ion concentration revealed in the range of altitudes 15000–20000 km, near the boundary of the earth's gas envelope" (Gringauz, 1961).

Actually two different theories which could qualitatively explain the existence of the break in the plasma density distribution with height, that occurs at an altitude of a few earth radii, (in current terminology – the plasmapause) were proposed in that same year, 1961.

Almost simultaneously with the publication of the experimental results obtained from the plasma-detectors on the lunar rockets, but independent from them, models of the magnetosphere were proposed in which there were convective motions of plasma in the external regions, such that an electric field would be produced in the equatorial plane directed across the magnetosphere (Dungey, 1961; Axford and Hines, 1961). In this model, a forbidden zone is formed close to the geomagnetic dipole, into which the plasma that takes part in the convection can not enter. The boundaries of this zone are formed by the 62° line of latitude which corresponds, in the equatorial plane, to a geocentric distance of $\sim 4.5 \, R_E$ (Axford and Hines, 1961). In this manner, the upper atmosphere, which is ionized by solar radiation, is found in the forbidden zone.

Of course, the laws relating the change in plasma density with geocentric height inside the forbidden zone, and outside of it, should be different.

In 1962, measurements of cold plasma, at geocentric distances up to $\sim 4\,R_E$ on high geomagnetically invariant latitudes ($\Lambda = 72°$), were made with instruments on board the Mars 1 rocket (Gringauz *et al.*, 1964). We shall return later to these observations – as far as is known, they have not been repeated for these particular heights and geomagnetic latitudes.

In 1963 Carpenter presented results showing electron-density distributions as a function of height, in the plane of the geomagnetic equator, which were obtained from analyses of ground-based observations of whistlers. In the $n_e(R)$ profiles, he found, at a geocentric distance of $\sim 4\,R_E$, a region where n_e decreases more rapidly and called this the 'knee' (Carpenter, 1963). The $n_e(R)$ profile from the whistler data is shown in Figure 1, which is taken from Carpenter's first report. The circles in this figure show the data obtained from detectors on Luna 2. The author remarked that the two sets of data were in agreement.

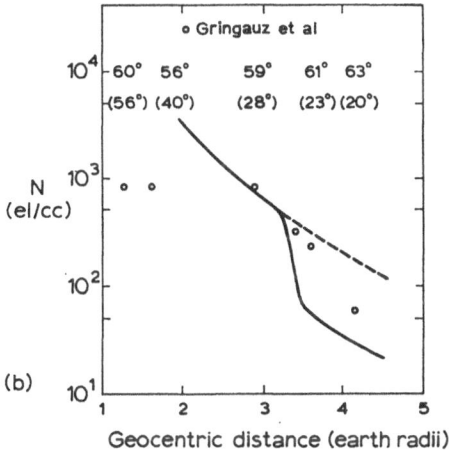

Fig. 1.   Charged particles density versus geocentric distance from data obtained by means of the ion traps on Luna 2 and from whistler data (Carpenter, 1963).

This report formed an important step in the study of the earth's plasma envelope, and it proved to be the beginning of a series of continuous studies by Carpenter himself, and by others. The results of these studies explained for the first time a number of peculiarities of the global structure and dynamics of the plasmasphere (continuous measurements of the height of the plasmapause in the equatorial zone, reactions of the plasmasphere to changes in geomagnetic activity, etc.).

The studies of the plasmasphere from whistler data are described in sufficient detail in Helliwell's review article (Helliwell, 1972), and so, here they will be mentioned only in so far as necessary to make comparisons to the data obtained by direct measurement.

In the present review, the data obtained from the various spacecraft are used as indicated in Table I.

TABLE I

Spacecraft data used

Spacecraft	Date of launching	Perigee (km)	Apogee (km)	Inclination to equator	Devices[a]
Luna 1	2. 1.59				Charged-particle traps
Luna 2	12. 9.59				Charged-particle traps
Mars 1	1.11.62				Charged-particle traps
Electron 2	30. 1.64	460	68 200	61°	Charged-particle trap
Electron 4	11. 7.64	460	66 235	61°	Charged-particle trap
OGO 1	5. 9.64	280	150 000	31°	Charged-particle trap of modulation type (Faraday cup), ion mass-spectrometer
IMP 2	4.10.64	197	93 910	33°	Charged-particle trap of modulation type (Faraday cup), charged-particle trap-retarding potential analyzer
OGO 2	14.10.64	415	1525	88°	Ion mass-spectrometer
OGO 3	7. 6.66	295	122 000	31°	Ion mass-spectrometer, charged-particle trap of modulation type, cylindrical electrostatic analyzer
OGO 5	4. 3.68	282	145 500	31°	Ion mass-spectrometer, Langmuir probe

[a] only devices data of which are used in this review are mentioned.

2. The Plasmapause at Low Geomagnetic Latitudes and its Variations

2.1. GENERAL MORPHOLOGY

The year 1964 proved to be very fruitful for further plasmaspheric studies, for, within that year, 4 satellites were launched into eccentric earth orbits with high apogees, carrying on board instruments to measure very low-energy charged particles. These satellites were the USSR Electron 2 and Electron 4, and the USA IMP 2 and OGO 1 (see Table I).

Figure 2 shows the first results of the ion density distribution measurements made with charged-particle traps on board Electron 2, in 1964, and this figure clearly shows the region of sharply decreasing ion density – the plasmapause (Bezrukikh and Gringauz, 1965). In the results obtained from OGO 1, using a radio-frequency mass-spectrometer, the plasmasphere was clearly observed, almost simultaneously in both the proton and helium components (Taylor *et al.*, 1965). Because of the fact that the inclination of the OGO 1 orbit to the equator and, as far as is known, the inclination to the equator of other satellites having highly eccentric orbit and having instruments (for studying the plasmasphere) is $\sim 31°$, all data obtained from this satellite refer to geomagnetic latitudes of $\leqslant 45°$.

Using the data obtained by direct measurements made on satellites whose orbits are inclined to the equator, it is possible to determine the position of the plasmapause in a meridional plane. Generalizing the ground-based observations of whistlers, and

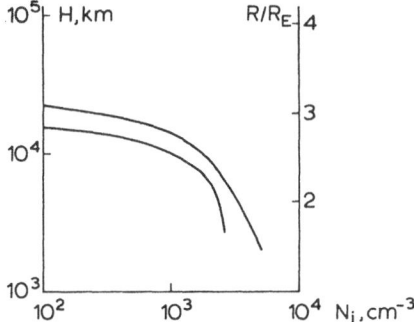

Fig. 2.   Distribution of ion density with height obtained by means of the charged particle trap on board Electron 2 satellite. The upper curve corresponds to 2.2.1964, the lower to 2.13.1964 (Bezrukikh and Gringauz, 1965).

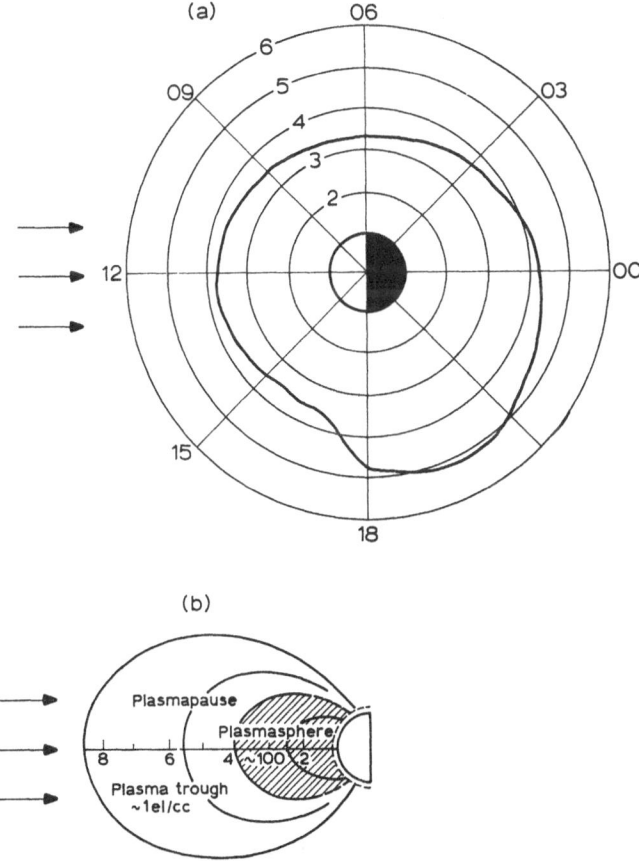

Fig. 3.   Model of the plasmapause for moderate geomagnetic activity (a) equatorial cross-sections; (b) meridional cross-section (LT ~ 14 h) (Carpenter, 1966).

ion measurements obtained by Taylor *et al.* from the OGO 1 satellite, Carpenter (1966) published a model of the plasmasphere, shown in Figures 3a and 3b.

Figure 3a shows the dependence of the geocentric distance of the plasmapause in the equatorial plane on local time, corresponding to an average value of geomagnetic activity. A notable feature of this diagram is its asymmetry, caused by the presence of the bulge in the evening region.

When geomagnetic activity increases, the plasmasphere is compressed, and its asymmetry increases; and when it decreases, the plasmasphere moves away from the earth in the equatorial plane to a distance of $R=6\ R_E$, and the asymmetry decreases (Carpenter, 1966).

Figure 3b shows a meridional cross-section of the daytime side of the plasmasphere (~ 14h LT). The shaded region is bounded by the magnetic shell $L=4$ and has an electron density of $n_e \sim 100\ \mathrm{cm}^{-3}$ which, close to the plasmapause, and $\sim 1\ \mathrm{cm}^{-3}$ outside. For geomagnetic latitudes $>45°$, the cross-section of the plasmasphere is indicated by a dashed line, because the author of the model lacked observational data for this region.

On the IMP 2 satellite, observations of the cold plasma were made with charged-particle traps using retarding potentials (Serbu and Maier, 1966), and with traps of the modulation type (Binsack, 1967). The authors of the former experiment, after reducing the data, did not find the region of sharply decreasing plasma density – the plasmapause; whereas in the latter experiment, it was observed. As far as we know, the results of investigation of the causes of why the data of the former experiment contradict the numerous results of other authors, has not yet been published.

To evaluate the reliability of the measurements of n_i, data on ion and electron densities at geocentric distances of $\sim 2\ R_E$ were obtained from five independent experiments and compared (whistler data and data obtained from the Electron 2 and OGO 1 satellites were included). In all cases, the density was found to lie between the limits $2 \times 10^3\ \mathrm{cm}^{-3}$ and $4 \times 10^3\ \mathrm{cm}^{-3}$ (the data from Electron 2 and OGO 1 were taken at a time near the minimum of the solar activity cycle) (Brinton *et al.*, 1968).

Figure 4 shows the distributions of n_i as functions of the McIlwain L parameter obtained using a mass-spectrometer at the time of maximum of solar activity, in March 1968, during the first 4 revolutions of the satellite OGO 5 around the Earth (Harris *et al.*, 1970). In all cases of $L=2$, n_i has the value of a few thousands per cm^3, and reaches, in one case (profile 7) a value of $\sim 10^4\ \mathrm{cm}^{-3}$.

In six out of the eight profiles, the plasmapause is clearly visible (the exceptions being profiles 1 and 4). The somewhat unusual $n_i(L)$ profile number 6 is interesting for it has two regions of sharply decreasing n_i (for $L=4$ and for $L=5.8$). In some of the $n_i(L)$ profiles, oscillations were observed. The authors of this paper referred to a private communication from P. Coleman, noting that simultaneous measurements made on the same satellite showed that there were oscillations in the magnetic field. Therefore the authors considered these results as evidence of some spatial fluctuations (in L). It should be noted here that, with data from only one spacecraft, it is not possible to determine whether the fluctuations are spatial or temporal.

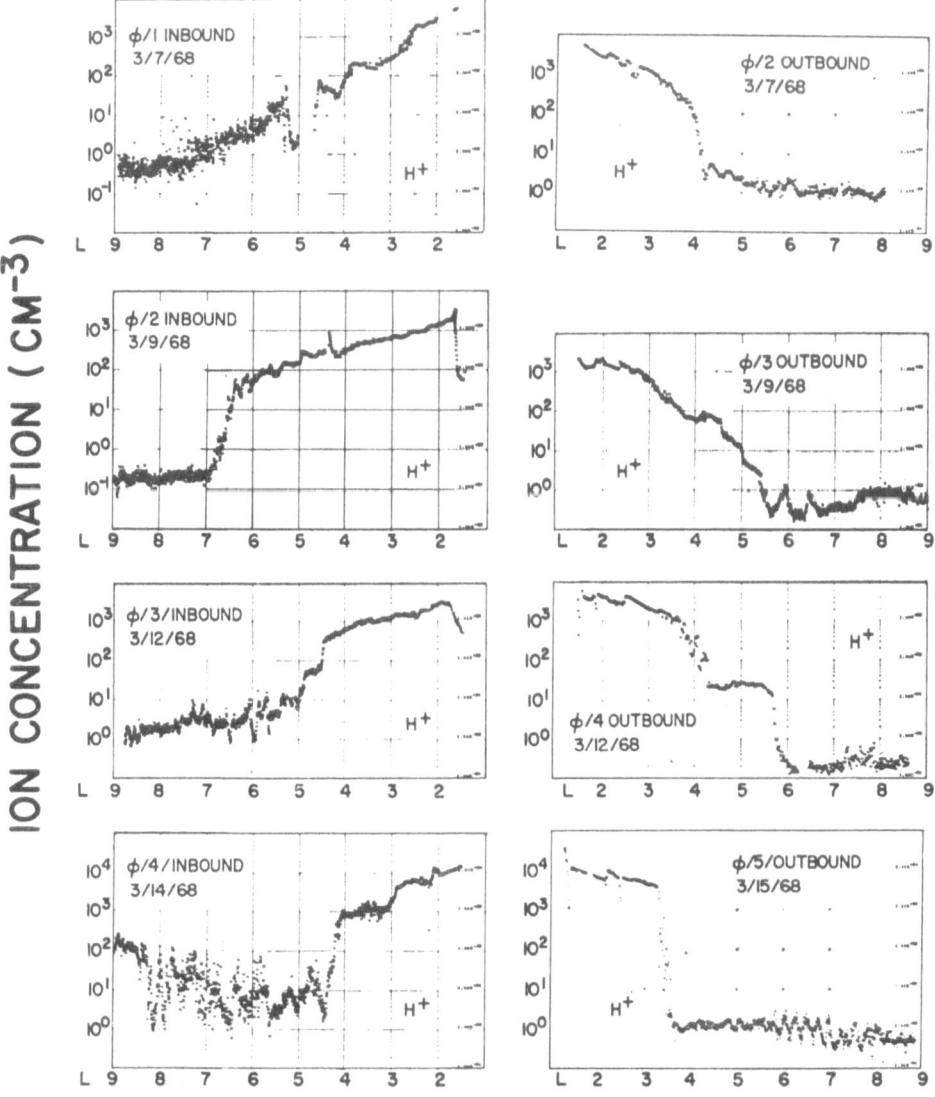

ION CONCENTRATION (CM^{-3})

Fig. 4. H^+ ion density distribution [$n_i(L)$ profiles] from ion mass-spectrometer data obtained on board the OGO 5 satellite (Harris *et al.*, 1970).

From the $n_i(L)$ profiles of Figure 4, it is evident that the position of the plasmapause changes within wide limits from $L \sim 3.5$ (profile 8) to $L = 6.5$ (profile 3).

2.2. DEPENDENCE OF PLASMAPAUSE POSITION ON GEOMAGNETIC ACTIVITY

Figure 5 shows the dependence of the L coordinate of the plasmapause on the maximum K_p index of each day, preceding measurements obtained in 1964 from the Electron 2 and Electron 4 satellites (Bezrukikh, 1968, Figure 5a), in 1966 from OGO 3

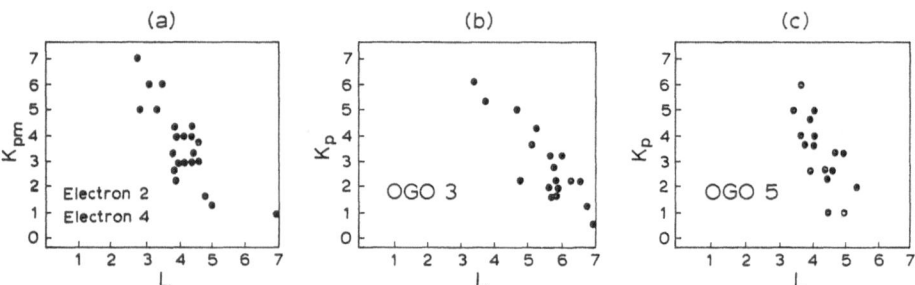

Fig. 5. The relation between L coordinate of the plasmapause (determined from ion density measurements) and K_p index averaged over the preceding 24 h: (a) 1964 – from Electron 2 and Electron 4 data (Bezrukikh, 1968); (b) 1966 – from OGO 3 data (Taylor et al., 1968a), (c) 1968 – from OGO 5 data (Chappell et al., 1970).

(Taylor et al., 1968a, Figure 5b), and in 1968 from OGO 5 (Chappell et al., 1970, Figure 5c).

From the results of observations of the average (not taking into account the local time) position of the plasmapause obtained from the IMP 2 satellite (1964), the following empirical formula for determining the position L, was proposed:

$$L = 6 - 0.6\,K_p \tag{1}$$

where the K_p index to be used corresponds to the time of measurement (Binsack, 1967).

Other authors prefer to use, for comparison, the maximum K_p index for the day preceding the measurement, obviously because of the inertia of the plasmasphere due to the low velocities with which the cold plasma in the magnetosphere moves. Together with this fact, one of the reasons for using the maximum K_p from the preceding day in the drawing of Figure 5, is a historical one – the desire of authors to compare satellite data with the analogous data obtained at first from whistler studies (Carpenter, 1966).

The question of the rate of reaction of the plasmasphere to changes in magnetic activity at various local times was discussed in a paper by Chappell et al., 1970, where they note that the time delay between changes in the geomagnetic activity level and in the plasmasphere position, of 2–6 h agree well in the region of 02 h local time, but for the region of 10 h LT, the agreement of temporary changes of 6, 12, or 24 h do not lead to significantly different results. Around 10 h LT, the drop in density close to the plasmapause, that accompanies an increase in geomagnetic activity, become far sharper. The authors' remark on the advisability of looking for a new index that would characterize the convection of the magnetospheric plasma, and that would be more suitable than the K_p index for comparisons of the type shown in Figure 5.

On the OGO 5 satellite, as well as the ion mass-spectrometer, there was a spherical Langmuir probe, with 6 cm diameter, with which measurements of the electron component of the plasma were made (Freeman et al., 1970). During the flight, the probe was in the shadow of the solar cell panels, for part of the time, and so its measurements were not influenced by photoemission from the surface of the probe (although it is possible that photoelectrons from the illuminated part of the spacecraft could fall into it). The experiment was designed to measure electron densities $\geqslant 10^{-3}$ cm^{-3} (for this magnitude of n_e the Debye radius becomes comparable to the spacecraft dimensions, that is 2.5 m). When the spacecraft is illuminated by the sun, and is in the plasmasphere the photoemission from the spacecraft is considered to be equal to the photoemissive current in the interplanetary medium, and thus is subtracted from the measurements made in the plasmasphere.

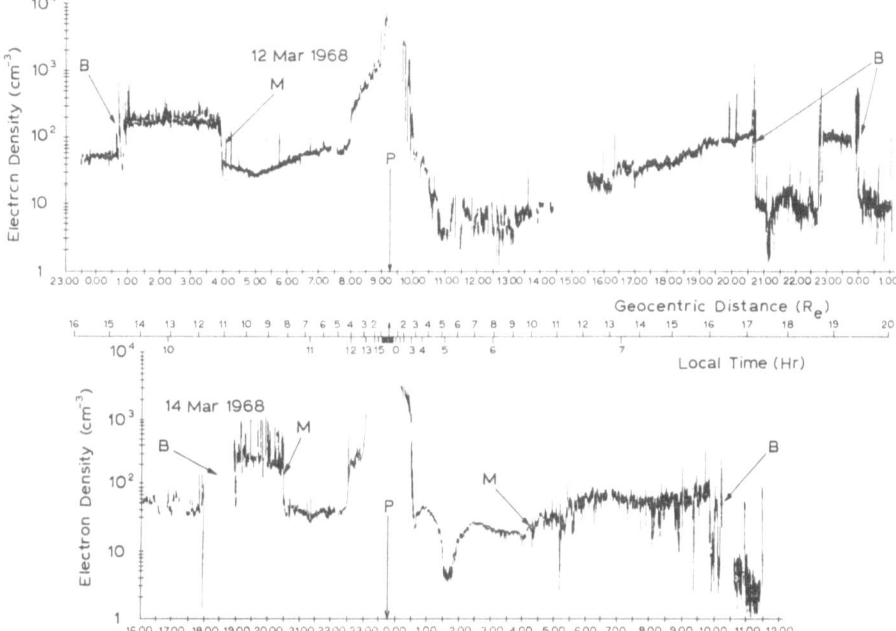

Fig. 6.   Samples of $n_e(R)$ profiles obtained by means of a spherical Langmuir probe on board OGO 5 satellite 12 and 14 March, 1968. The arrows marked by the letters B and M indicate the positions of the shock-wave front and the magnetopause respectively; the letter P marks the position of perigee (Freeman *et al.*, 1970).

Figure 6 shows sample $n_e(R)$ profiles, drawn by a computer, with no additional reducing. The arrows marked with the letters B and M indicate the positions that correspond to the shock-wave front, and the magnetopause, respectively, and the letter P marks the position of perigee. These drawings clearly show that the use of Langmuir probes succeeded for the first time in directly detecting the plasmapause in the electron component of the plasmasphere. The data presented in Figure 6 obtained simultaneously with $n_i(L)$ profiles nos. 5, 6, 7, and 8 in Figure 4. If one takes into account the fact that for OGO 5 the geocentric distance in earth radii is only a little smaller than the L coordinate, then the agreement between the position of the plasmapause, as determined from the n_e data from the Langmuir probe, and from the n_i data from the ion mass-spectrometer, may be considered satisfactory.

2.3. THE DIURNAL BULGE

An important characteristic of the plasmapause in the equatorial cross-section of the plasmasphere is the bulge in the evening sector, that was discovered in whistler studies and was shown in Figure 3a.

The bulge was also observed in the results obtained from traps on board the Electron 2 and Electron 4 satellites, in which the magnitudes of the L coordinates of the plasmapause were systematically higher in the evening region than in the morning region (Bezrukikh, 1968). It was also observed on the OGO 1 and OGO 3 satellites (Brinton *et al.*, 1968).

The OGO 5 satellite passed through the evening sector more than 30 times encountering various levels of geomagnetic activity. Using the ion mass-spectrometer on board, a number of detailed $n_i(L)$ profiles, close to the boundaries of the bulge, were obtained, from which the authors of the measurements drew a number of conclusions about the structure and dynamics of this region of the plasmasphere (Chappell *et al.*, 1969).

First of all, the authors noted that, for different levels of geomagnetic activity, the bulge, on the average, is symmetrical about the meridian LT = 18 h. This result agrees with Figure 7, which shows the location of the plasmapause, encountered by the satellite in the vicinity of the 18 h meridian, for the various K_p indexes. The authors of the ion mass-spectrometer experiment consider that the results of their measurements agree well with a picture of the phenomena in the region of the bulge, that may briefly be described as follows: as the magnetic activity decreases, the electric

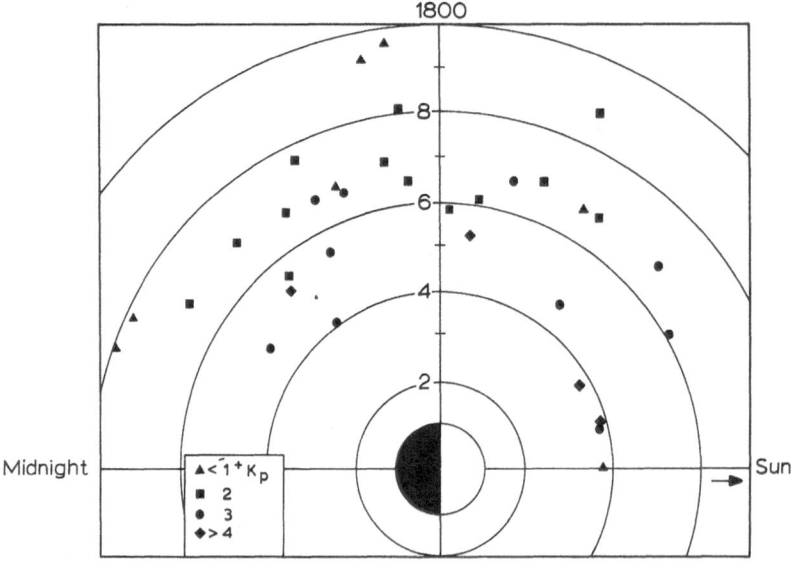

Fig. 7. Locations of the plasmapause in the vicinity of the bulge in the dusk sector of the magnetosphere for the various K_p indexes from the ion mass-spectrometer data taken during 34 passes of the OGO 5 satellite (Chappell *et al.*, 1969).

field across the equatorial section of the magnetosphere (in the direction morning to evening) decreases, the forbidden zone increases, and the bulge becomes filled with plasma from the lower region of the ionosphere.

When the magnetic activity increases, the convection electric field increases and the forbidden zone decreases (the plasmapause moves closer to the earth); in this case, part of the plasma from the bulge can detach from the plasmasphere and take part in the convection, moving towards the day-region of the magnetosphere. Figure 8 shows the $n_i(L)$ profile, corresponding to the passage of OGO 5 through the bulge on October 1, 1968 at 13 h–16 h LT. The plasma detected in the intervals $5.5 < L < 7.7$,

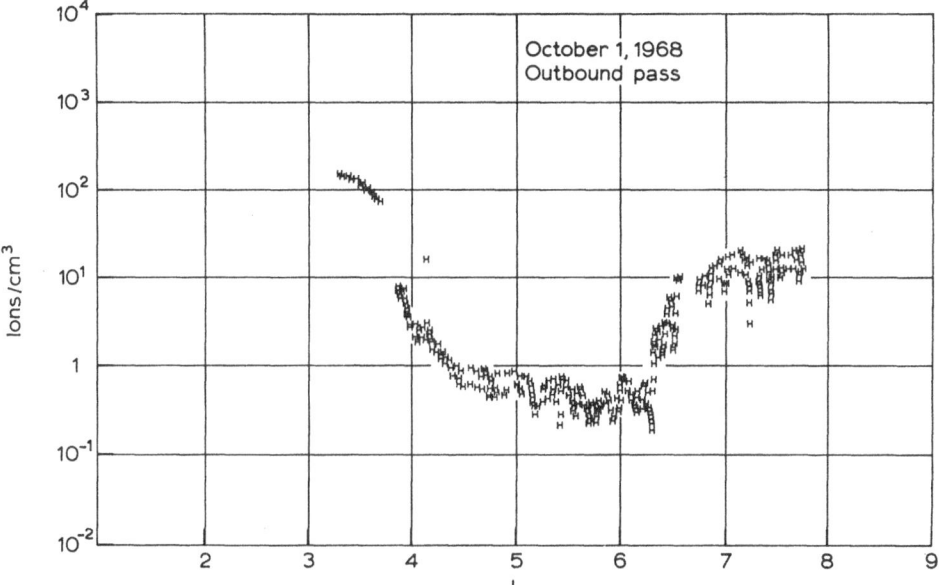

Fig. 8. $n_i(L)$ profile obtained by means of the ion mass-spectrometer on board OGO 5 satellite during its passage through the bulge 10.1.1968 (Chappell *et al.*, 1969). After passing through the plasmapause the satellite registered a plasma cloud separated from the bulge.

is interpreted as separated from the plasmasphere. Before these measurements were taken, the magnetic activity had been low for about one day and then, within ~ 15 h, it became high. In such a period of time, the total amount of plasma in the plasmasphere decreases.

This process is illustrated in Figure 9. This figure shows the separation of a cloud of plasma from the bulge, which, on entry into the convecting magnetospheric plasma, moves toward the day region of the magnetopause, then along the magnetopause to the tail of the magnetosphere, where it enters into open field lines and goes off into interplanetary space.

The results of the $n_i(L)$ profile measurements, made with the OGO 5 mass-spectrometer, do not, as the authors remarked, support the suggestion made by

Nishida that the bulge may be considered as 'new' plasma having a non-ionospheric origin), which may exhibit an eddy-type flow (Nishida, 1966).

The conclusion drawn by Chappell *et al.* on the origin, the structure and the mechanism for growth and diminution of the bulge, and their interpretation of experimental data of the type shown in Figure 8, are very interesting and important, although it is possibly still too early to consider them final. Specifically, some concern is caused by the contradiction between two sets of data relating the characteristics of the bulge and magnetic activity. One set, discussed above, was that obtained on OGO 5 which showed that the average symmetry of the bulge around the 18 h meridian was independent of magnetic activity. The second, (reproduced in the preprint of Chappell *et al.*, 1969) was a diagram showing changes of the bulge during changes in magnetic activity that was made by Carpenter from whistler data (Figure 10), in which an asymmetry of the bulge around LT = 18 h is obvious.

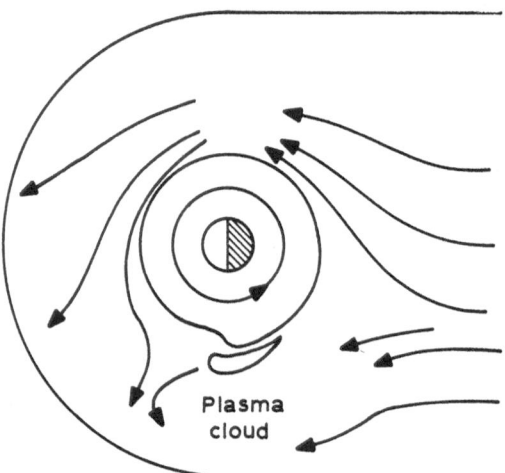

Fig. 9.   A possible scheme of motion of plasma cloud detached from the bulge in the equatorial plane of the magnetosphere.

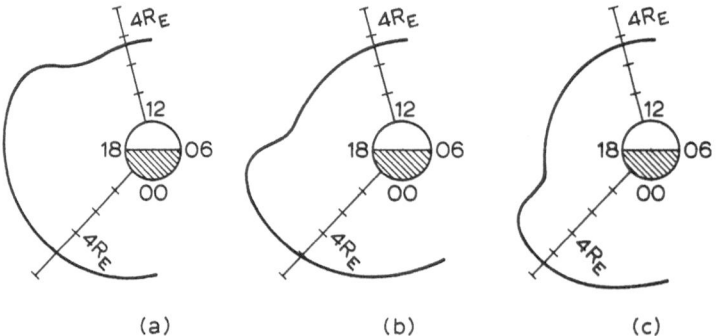

Fig. 10.   Form of the bulge (after Carpenter) corresponding to different levels of geomagnetic activity (a) high, (b) moderate, (c) low (Chappell *et al.*, 1969).

One must keep in mind the fact that the OGO 5 satellite crossed the bulge once every 2.5 d, whereas the whistler studies measure the position of the plasmapause almost continuously. Therefore, to resolve this contradiction, a far better statistical sample of satellite data is required, than that used, so far, in the paper by Chappell *et al.* (1969).

2.4. CHEMICAL COMPOSITION OF THE PLASMASPHERE

According to the data obtained from the radio-frequency mass-spectrometer carried on board OGO 1, the density of He^+ ions in the plasmasphere is 1% of the H^+ ion density. The plasmapause is found to be at practically the same heights for both components (Taylor *et al.*, 1965).

Figure 11a shows the $n_i(L)$ profiles obtained from the OGO 3 satellite (Taylor *et al.*, 1968a). One can see from this figure that the results obtained from OGO 3 are in complete agreement with data from OGO 1.

Figure 11b shows the $n_i(L)$ profiles obtained with the magnetically deflecting ion mass-spectrometer on OGO 5 (Harris *et al.*, 1970). Besides the He^+ and H^+ ion distributions, the $n_i(L)$ profile for O^+ ions is also shown. These OGO 5 measurements confirmed the fact that the He^+ ion density is $\sim 1\%$ of the proton density, and showed that the O^+ ion density is $\sim 0.3\%$ of the proton density.

3. Polar Ionosphere and Plasmasphere

According to the idealized model of the meridional cross-section of the plasmasphere, shown in Figure 3b (Carpenter, 1966), beyond the boundaries of the geomagnetic shell that passes through the plasmapause in the equatorial plane, density of cold plasma is very small at all latitudes at sufficiently large heights (>1000 km). In the construction of this model as mentioned above, direct measurements were used (from the ion mass-spectrometer on the OGO 1 satellite). Further data supporting this model were obtained from a number of other experiments. On the Alouette 1 and Alouette 2 satellites, which had orbits with low heights (<3000 km), the plasmapause was detected by the sudden disappearance or diminution of whistlers propagated from the other hemisphere (Carpenter *et al.*, 1968). For values of $L\sim 4$, the plasmapause was found simultaneously by ion mass-spectrometers and wide-band receivers on the OGO 1 and OGO 3 satellites, and also by whistler reception at Antarctic stations (Carpenter *et al.*, 1969). However, although measurements made on the Electron 2 and Electron 4 satellites with charged-particle traps in the vicinity of the equatorial plane were in good agreement with all details of the whistler observational results, the agreement was significantly worse for the $n_i(L)$ profiles, obtained at higher latitudes (Bezrukikh, 1968), on which the plasmapause was absent in a number of cases. Figure 12 shows these $n_i(L)$ profiles obtained from Electron 4 for various levels of magnetic activity. In comparison with the $n_i(L)$ profiles from the OGO satellites, these show a comparatively high value of L for $R\sim(2-4)R_E$. The decrease in n_i with growth of R or L is monotonic and the magnitude of n_i is rather large far

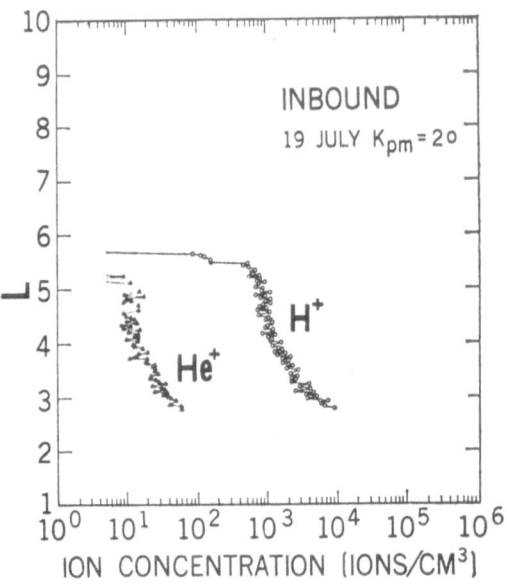

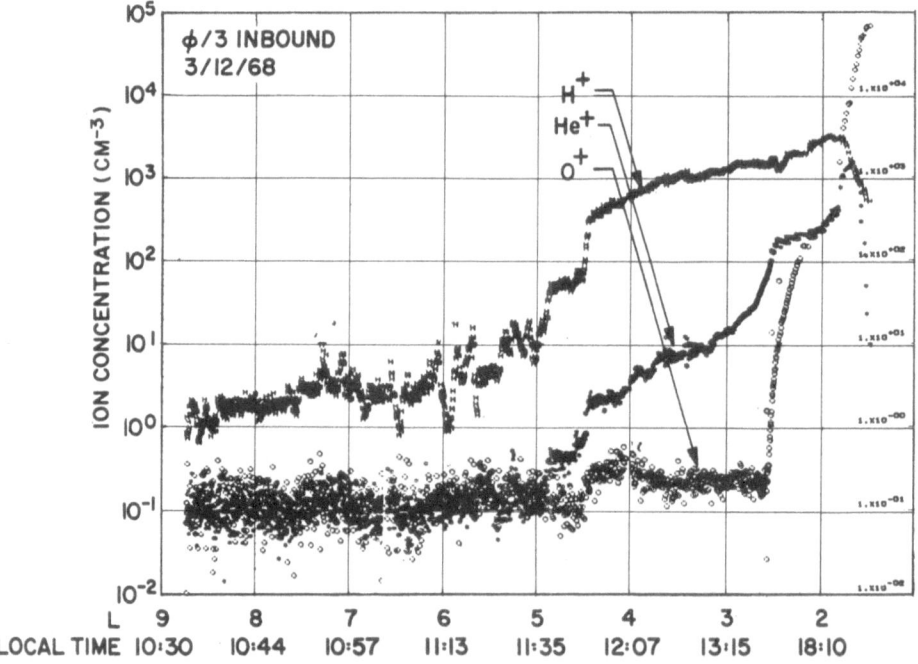

Fig. 11. n_i (L) profiles: (a) for H⁺ and He⁺ ions from data obtained on board OGO 3 satellite (Taylor *et al.* 1968a); (b) for H⁺, He⁺ and O⁺ ions from data obtained on board OGO 5 satellite. (Harris *et al.*, 1970).

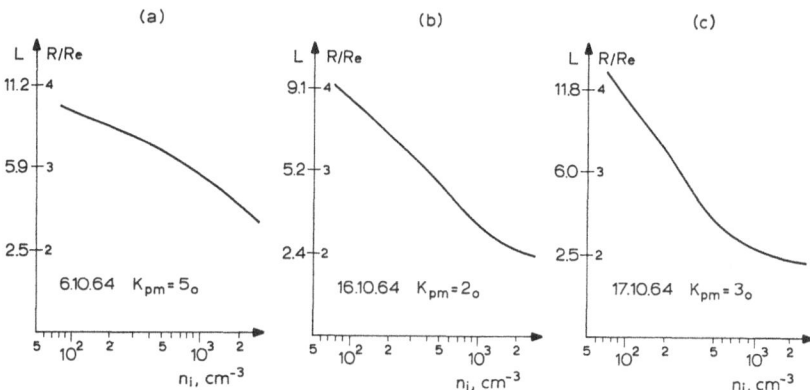

Fig. 12. $n_i(L)$ profiles obtained on board Electron 4 satellite (Bezrukikh, 1968).

beyond the geomagnetic shell, when the plasmapause position should have been given by, for example, Equation (1) above.

A few remarks will now be made on some of the characteristics of the polar ionosphere and the polar wind (although the latter is discussed elsewhere in this volume (Mange, 1972). From a number of satellites with polar orbits at heights of up to 3000 km, and using probes, traps, and ion mass-spectrometers as well as radio methods, it was discovered that, in the upper part of the F layer of the ionosphere, at geomagnetic latitudes of $\Lambda > 60°$, the density of charged particles sharply decreases, in comparison to the density at the same heights at low and middle latitudes, and that the ion composition changes, there being a sharp decrease in the concentrations of the light ions H^+ and He^+ (Bowen et al., 1964; Brace and Reddy, 1965; Taylor et al., 1968b, Taylor et al., 1969; Barrington et al., 1966; Thomas et al., 1966; Hagg, 1967). A number of authors (Bauer, 1966; Dessler and Michel, 1966) explained this phenomenon by the escape of light ions from the polar regions along open field lines; the most complete and well-grounded form of this idea is found in the papers on the polar wind by Axford (1968) and by Banks and Holzer (1968, 1969). The plasmapause (particularly during high magnetic activity) is situated at lower latitudes than the boundaries of the stable zone of trapped radiation (i.e. $L \sim 8$–9, see Vernov et al., 1969) within which the magnetic lines of force are obviously closed. The formation of the plasmapause may be connected with the polar wind (Axford, 1969) by a hypothesis on the convection of lines of force that become opened during part of the daily convection cycle (Nishida, 1966).

In connection with the polar wind theory, we should like to turn attention to the following peculiarity of the polar ionosphere: amongst the results obtained in 1965 with the ion mass-spectrometer on OGO 2 at heights of 450–1500 km, besides the decrease in ion density to $\sim 10^2$ cm^{-3} and the depletion of light ions at latitudes of $> 60°$, a changing though frequently existing peak was observed at $\gtrsim 80°$, in which the ion density increased by two orders of magnitude above the value at lower latitudes. This peak of ionization often has a rather high number of H^+ ions, and its electron component is systematically observed with radio methods (Thomas *et al.*, 1966). The lines of force corresponding to this polar peak of ionization are obviously open; the formation of this peak has evidently to be considered (particularly from the point of view of the polar wind). To what heights does this polar peak reach? It is difficult to answer this question, because direct measurements of the plasma in the polar region have not been made for heights above 3000 km.

The only spacecraft carrying instruments to measure the cold plasma, that passed, in 1962, through the regions of invariant geomagnetic latitudes $\Lambda \leqslant 72°$, at a geocentric distances of a few R_E was, as far as is known to us, the Soviet Mars 1 (Gringauz *et al.*, 1964). It registered significant ion densities in regions where, according to the model shown in Figure 3b, such densities ought not to exist.

Apparently, the phenomena related to cold-plasma distributions at great heights close to the equatorial plane have been studied far better than those at higher latitudes.

In concluding their article on simultaneous observations of the plasmapause made with the OGO satellite and at the Antarctic stations, Carpenter *et al.* (1969) wrote: "The results presented above may be deceptive in their simplicity. Data not presented here indicate that the plasmapause is extremely complex, with regions of irregular behavior, periods of rapid expansion or compression and variations in details of the plasma profile at the boundary. Further generation of correlation studies are needed to obtain a proper description of these effects."

It is necessary to keep these words in mind when using simple models of the plasmasphere, particularly when using them at comparatively high geomagnetic latitudes.

4. Energy of the Particles that Populate the Plasmasphere

4.1. INTRODUCTION

Charged particles with a great range of energies coexist in the plasmasphere, and it is only possible to study their energies using direct methods: as was already mentioned, observations of whistlers do not give any information on the energy of the particles.

The basic population of the plasmasphere consists of particles where energies do not exceed 1 eV by more than order of magnitude. This was shown in the experiments on the first lunar spacecraft (Gringauz *et al.*, 1960a). The particles in the radiation belts were also observed on these spacecraft (Vernov *et al.*, 1959) and a comparison of the results showed that the boundary of the plasmasphere (plasmapause in current terminology) was located inside the outer radiation belt, and that therefore, there were particles with subrelativistic and relativistic velocities in the regions under consideration as well as the cold-plasma particles.

The observations at comparatively low values of L of intense fluxes of electrons, with energies ranging from hundreds of eV to tens of keV (Frank, 1966), filled the gap in the energy distribution of charged particles in the plasmasphere between the 'cold' particles, and the high-energy particles of the radiation belts, and showed that particles with almost all energies, ranging from ~ 1 eV to hundreds of MeV, coexist in the plasmasphere.

The movement of these particles in the region close to the plasmapause, depends on their energies; the cold-plasma particles, because of their low velocities and corresponding small magnetic moment, do not drift due to the magnetic field gradient, but strongly react to the convective electric field; the charged particles of the radiation belts, whose motion is determined by the gradient drift to a considerable degree, react far less strongly to the electric field. The boundary of the forbidden zone formed by the convective electric field for the cold magnetospheric plasma is no obstacle for these high-energy particles. This does not mean that long-term effects of the electric field on the motions of the particles in the radiations belts can be neglected. For the fluxes of particles with intermediate energies ($E < 100$ keV) the convective electric field can cause a considerable asymmetry in the east-west direction (Axford, 1969; Tverskoy, 1970).

As was shown in the experiment carried out with electrostatic analyzers on the OGO 3 satellite (Schield and Frank, 1969), the plasmapause may be observed, not only in measurements of particles with energies of ~ 1 eV but also in measurements of electrons with energies of the order of hundreds of eV. We shall return to this later.

4.2. Cold plasma temperature

The number of publications of experimental determination of the cold plasma temperature is small.

By comparing the ion currents from the four traps, which were mounted on each of the Luna 1 and Luna 2 satellite, in 1959, and which had different potentials on the outer grids, the ion temperature T_i of the cold plasma that surrounds the Earth was first defined roughly and it was determined that does not exceed a few tens of thousands of degrees (Gringauz et al., 1960; Gringauz, 1961). Later publications, which contained experimental data for charged-particle temperature at geocentric distances of up to a few R_E, were based on measurements obtained using charged-particle traps with retarding potentials on the IMP 2 satellite (Serbu and Maier, 1966, 1967). These articles were critized a number of times because the authors, after reducing the first results, did not find the plasmapause (Gringauz, 1967; Binsack, 1967; Harris et al., 1970).

According to the conclusions of Serbu and Maier, at a distance of $\sim 4 R_E$, $T_e \sim 1$ eV; T_e increases with height according to the law $\sim R^2$, and T_e is considerably smaller than T_i. Because of the peculiarity of these results, which was pointed out (the absence of the plasmapause), it is difficult to determine which of the temperature data, obtained on IMP 2, refers to the plasmasphere, and which refers to the plasma outside of the plasmasphere. As was noted by Bezrukikh et al. (1967), an explanation of the causes for having $T_i \gg T_e$, is very difficult to find.

A method for determining T_i in the plasmasphere has been suggested, using current variations in a trap with a zero potential outer grid produced by spacecraft rotation (Gringauz et al., 1967). Because of the insufficient amount of data obtained from the trap on Electron 2, only the upper limits of T_i were calculated, which for heights of $\leqslant 10000$ km for two passes of the satellite, turned out to be $7000°$ and $10000°$, respectively (Gringauz et al., 1967).

At a geocentric distance of $\sim 4.2\ R_E$, the upper limit of T_i was determined to be 1.1 eV (Bezrukikh et al., 1967). These estimates were made before data on the orientation of the trap relative to the spacecraft velocity vector were obtained. Subsequently, using data from the solar orientation system of the satellite and the magnetometer, the orientation of the trap on Electron 2 relative to its velocity vector was determined, and corrections were made to account for some effects that had not been taken into account in the earlier estimation of the upper limit of T_i. This resulted in a lowering of the earlier estimate for the magnitude of T_i, from the Electron 2 data, by a factor of two.

The determination of electron temperature T_e, from the spherical Langmuir probe data from OGO 5, was complicated by photoemission. The authors of the experiment gave values for the magnitude ranging from 3×10^3 K at perigee (at a height of 300 km) to 3.5×10^4 K (close to the magnetopause). In calculating n_e inside the plasmasphere using the value $T_e = 10000$ K the experimenters estimated the error as being not greater than 50% (Freeman et al., 1970).

4.3. Electrons with energies $100\ \mathrm{eV} \leqslant E \leqslant 50\ \mathrm{keV}$ in the plasmasphere and on its boundaries

The possibility of detecting the plasmapause in electron energy distribution at energies of $100\ \mathrm{eV} \leqslant E \leqslant 700\ \mathrm{eV}$ was demonstrated with the data from the electrostatic analyzer on the OGO 3 satellite (Schield and Frank, 1969).

Figure 13a shows measurements, made on the inbound pass of OGO 3 on 21 June, 1966, of the electron currents with the following energies: 80–100 eV curve no. (3), 190–330 eV (4), 310–540 eV (5), 410–720 eV (6), 640–1100 eV (7), and 990–1700 eV (8).

Figure 13b shows $n_e(L)$ profiles, constructed from Schield and Frank's data which were obtained from the same satellite pass, for electrons with energies in the range $100\ \mathrm{eV} \leqslant E \leqslant 700\ \mathrm{eV}$ (dotted curve), and in the range $700\ \mathrm{eV} \leqslant E \leqslant 50\ \mathrm{keV}$ (continuous curve).

The location of the peaks on curves (3)–(6), Figure 13a, and on the dotted curve on Figure 13b agree, to a good degree of accuracy, to the location of the plasmapause, as given by radio-frequency mass-spectrometer data that was obtained simultaneously on the same spacecraft. Schield and Frank note that the spatial distribution of electrons with energies of $100\ \mathrm{eV} \leqslant E \leqslant 700\ \mathrm{eV}$, is usually as shown in Figure 13.

There are cases, however, when the $n_e(L)$ profiles of electrons with energies in the given range do not show any change at the plasmapause. This usually occurs at times close to strong magnetic disturbances, when the region of trapped radiation becomes filled with low energy particles (with energies from a few hundred eV to tens of keV). This was, in particular, the case on 21 July, 1966.

The existence in the plasmasphere of rather sharp peaks in the density of electrons having energies in the range $100\ \mathrm{eV} \leqslant E \leqslant 700\ \mathrm{eV}$ indicated that the interaction of the magnetospheric plasma that takes part in the convective motion, with ionospheric (plasmaspheric) plasma that corotates with the earth, may be accompanied by processes which are related to local heating of plasma, and which produce the accelerated electrons that are observed to have energies up to 700 eV.

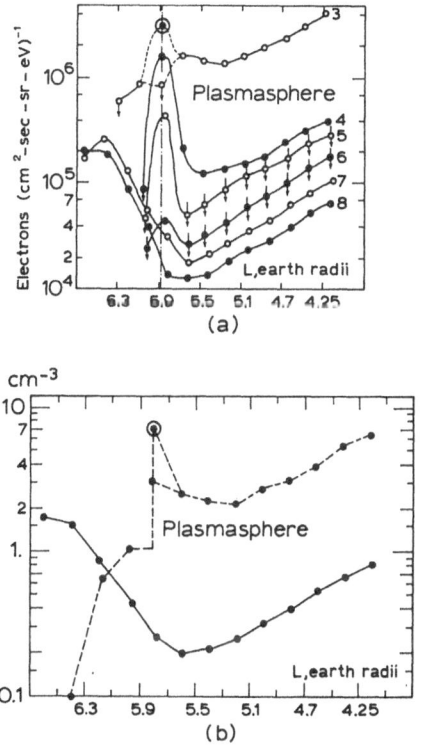

Fig. 13. (a) Electron fluxes with energies of 80–100 eV (3); 190–330 eV (4); 310–540 eV (5); 410–720 eV (6); 640–1100 eV (7); 990–1700 eV (8). (b) $n_e(L)$ profiles for electron energies of – – – – 100 eV $\leqslant$ $\leqslant E \leqslant 700$ eV, – – – – – 700 eV $\leqslant E \leqslant$ 50 keV (according to the data of Schield and Frank, 1969).

5. Conclusions

Although the first articles on the observation of the plasmapause did not attract the attention of many geophysicists (in the first six years from 1960 to 1965 the number of articles on the plasmasphere was less than 10) the situation has changed considerably in the past five years.

Theoretical papers have appeared, dealing specifically with the formation of the plasmapause (Block, 1966; Nishida, 1966; Samokhin, 1967a, 1967b, 1969; Brice, 1967; Mayr, 1968; Kavanagh et al., 1968; Raspopov, 1969a, 1969b).

The reduction of observational results of whistlers and direct plasmaspheric measurements from the Electron 2 and 4 satellites, and the OGO 1, 3 and 5 satellites have given a great wealth of information, and the results of the direct and indirect measurements are in good agreement in the equatorial plane (although at higher geomagnetic latitudes, the situation is less clear).

It has become generally agreed that the plasmapause is related to the important ionospheric and magnetospheric phenomena.

The propagation of very-low-frequency radio waves (Helliwell, 1969, 1972), the peculiarities of the polar ionosphere structure (Thomas *et al.*, 1966; Taylor *et al.*, 1968b), the heating of the night side of the F region of the ionosphere (Nagy *et al.*, 1968), the generation of micropulsations in the geomagnetic field (Troitskaya and Gul'elmi, 1969), the basic current systems in the ionosphere, large-scale electric fields and the convection of cold plasma in the magnetosphere (Axford, 1969) – all these phenomena have proved to be related in one way or another to the plasmasphere or plasmapause.

With the help of empirical relations connecting the K_p index to both the solar wind velocity, and to the plasmapause location, it is possible to determine roughly the convective electric field along the evening meridian (Vasyliunas, 1968b).

The measurements of the plasmapause made in the morning sector are most convenient for investigation of the large-scale electric field in the magnetosphere (Raspopov, 1969b).

Evidently, measurements of plasmapause position and the use of these results for the estimation of the electric field in the magnetosphere allows the theory of large-scale variations in the radiation belts to be defined more exactly (Tverskoy, 1972).

All this allows one to foretell, (with no danger of being mistaken), that the study of the plasmasphere, and in particular, the plasmapause, will be significantly widened in the near future.

This is all the more necessary because many of the characteristics of the processes taking place near the plasmapause are not clear; experimental data on the cold plasma close to the earth at $R \geqslant 1.4\ R_E$ over high latitude regions are very meagre and, in particular are entirely lacking above the polar peak of ionization.

References

Axford, W. I.: 1968, *J. Geophys. Res.* **73**, 6855.
Axford, W. I.: 1969, *Rev. Geophys.* **7**, 421.
Axford, W. I. and Hines, C. O.: 1961, *Can. J. Phys.* **39**, 1433.
Banks, P. M. and Holzer, T. E.: 1968, *J. Geophys. Res.* **73**, 6846.
Banks, P. M. and Holzer, T. E.: 1969, *J. Geophys. Res.* **74**, 6317.
Bauer, S. J.: 1966, *Ann. Geophys.* **22**, 247.
Barrington, R. E., Belrose, J. S., and Nelms, G. L.: 1966, in *Electron Density Profiles in Ionosphere and Exosphere* (ed. by J. Frihagen). North-Holland Publishing Co., Amsterdam, p. 387.
Bezrukikh, V. V.: 1968, Paper presented at International Symposium on the Physics of the Magnetosphere, Washington; *Kosmich. Issled.* [*Cosmic Res.*] **2**, 271, 1970.
Bezrukikh, V. V. and Gringauz, K. I.: 1965, in *Issledovaniya Kosmicheskogo Prostranstva*, 'Nauka', Moscow, p. 177.
Bezrukikh, V. V., Breus, T. K., and Gringauz, K. I.: 1967, *Kosmich. Issled.* [*Cosmic Res.*] **5**, 798.
Binsack, J. H.: 1967, *J. Geophys. Res.* **72**, 5231.
Block, J. P.: 1966, *J. Geophys. Res.* **71**, 855.
Bowen, P. J., Boyd, R. L., Raitt, W. I., and Wilmore, A. P.: 1964, *Proc. Roy. Soc.* **A281**, 504.

Brace, L. H. and Reddy, B. M.: 1965, *J. Geophys. Res.* **70**, 5783.

Brice, N. M.: 1967, *J. Geophys. Res.* **72**, 5193.

Brinton, H. C., Pickett, R. A., and Taylor, H. A.: 1968, *Planetary Space Sci.* **16**, 899.

Carpenter, D. L.: 1963, Paper presented to the 14th General Assembly of URSI, Tokyo.

Carpenter, D. L.: 1966, *J. Geophys. Res.* **71**, 693.

Carpenter, D. L., Walter, F., Barrington, R. E., and McEwen, D. J.: 1968, *J. Geophys. Res.* **73**, 5511.

Carpenter, D. L., Park, C. G., Taylor, H. J., and Brinton, H. C.: 1969, *J. Geophys. Res.* **74**, 1837.

Chappell, C. R., Harris, K. K., and Sharp, G. W.: 1969, 'The Morphology of the Bulge Region of the Plasmasphere', preprint, Lockheed Palo Alto Research Laboratory.

Chappell, C. R., Harris, K. K., and Sharp, G. W.: 1970, *J. Geophys. Res.* **75**, 50.

Dessler, A. J. and Michel, F. C.: 1966, *J. Geophys. Res.* **71**, 1421.

Dungey, J. M.: 1961, *Phys. Rev. Letters* **6**, 47.

Frank, L. A.: 1966, *J. Geophys. Res.* **71**, 4631.

Freeman, R., Norman, K., and Willmore, A. P.: 1970, in *Intercorrelated Satellite Observations* (ed. by V. Manno and D. E. Page), Reidel, Dordrecht, p. 524.

Geisler, J. E. and Bowhill, S. A.: 1965, *J. Atmospheric Terrest. Phys.* **27**, 457.

Gliddon, J. E. C.: 1966, Aeronomy Rept. 12, University of Illinois, Illinois, Urbana, June.

Gringauz, K. I.: 1961, *Space Res.* **2**, 574.

Gringauz, K. I.: 1967, in *Solar-Terrestrial Physics* (ed. by J. W. King and W. S. Newman), Academic press, New York, p. 341.

Gringauz, K. I.: 1969, *Rev. Geophys.* **7**, 339.

Gringauz, K. I., Bezrukikh, V. V., Ozerov, V. D., and Rybchinsky, R. Ye.: 1960a, *Dokl. Akad. Nauk, SSSR* **131**, 1301.

Gringauz, K. I., Kurt, V. G., Moroz, V. I., and Shklovsky, I. S.: 1960b, *Astron. Zh.* **37**, 716.

Gringauz, K. I., Bezrukikh, V. V., Musatov, L. S., Rybchinsky, R. Ye., and Sheronova, S. M.: 1964, *Space Res.* **4**, 621.

Gringauz, K. I., Bezrukikh, V. V., and Breus, T. K.: 1967, *Kosmich Issled.* [*Cosmic Res.*] **5**, 245.

Gurnett, D. A. and Brice, N. M.: 1966, *J. Geophys. Res.* **71**, 3639.

Hagg, E. L.: 1967, *Can. J. Phys.* **45**, 27.

Harris, K. K., Sharp, G. W., and Chappell, C. R.: 1970, *J. Geophys. Res.* **75**, 219.

Helliwell, R. A.: 1969, *Rev. Geophys.* **7**, 281.

Helliwell, R. A.: 1972, this volume, p. 165.

Kavanagh, L. D., Freeman, J. W., and Chen, A. J.: 1968, *J. Geophys. Res.* **73**, 5511.

Kurt, V. G.: 1967, *Kosmich. Issled.* **5**, 911.

Mange, P.: 1972, this volume, p. 68.

Mayr, H. G.: 1968, *Planetary Space Sci.* **16**, 1405.

Nagy, A. F., Bauer, P., and Fontheim, E. G.: 1968, *J. Geophys. Res.* **73**, 6259.

Nishida, A.: 1966, *J. Geophys. Res.* **71**, 5669.

Raspopov, O. M.: 1969a, in *Solnechno-zemnaya physica* **1**, 240 (ed. by Soviet 'Solzne-Zemla'), Moscow.

Raspopov, O. M.: 1969b, Paper presented to IAGA symposium, Madrid.

Samokhin, M. V.: 1967a, *Kosmich. Issled.* [*Cosmic Res.*] **5**, 376.

Samokhin, M. V.: 1967b, *Geomagnetizm i Aeronomiya*, **7**, 411.

Samokhin, M. V.: 1969, *Kosmich. Issled.* **7**, 611.

Schield, M. A. and Frank, L. A.: 1969, 'Electron Observations between the Inner Edge of the Plasma Sheet and the Plasmasphere', preprint, Department of Physics and Astronomy, University of Iowa.

Serbu, G. P. and Maier, E. J. R.: 1966, *J. Geophys. Res.* **71**, 3755.

Serbu, G. P. and Maier, E. J. R.: 1967, *Space Res.* **7**, 527.

Taylor, H. A., Jr., Brinton, H. C., and Smith, C. R.: 1965, *J. Geophys. Res.* **70**, 5769.

Taylor, H. A., Jr., Brinton, H. C., and Pharo, M. W.: 1968a, *J. Geophys. Res.* **73**, 961.

Taylor, H. A., Jr., Brinton, H. C., Pharo, M. W., and Rahman, N. K.: 1968b, *J. Geophys. Res.* **73**, 5521.

Taylor, H. A., Jr., Brinton, H. C., Carpenter, D. L., Bonner, F. M., and Heyborne, R. L.: 1969, *J. Geophys. Res.* **74**, 3517.

Thomas, J. O., Rycroft, M. J., Colin, L., and Chan, K. L.: 1966, *Proc. NATO Advanced Study, Norway* (ed. by J. Frihagen), North-Holland Publishing Co., Amsterdam.

Troitskaya, V. A. and Gul'elmi, A. V.: 1969, *Uspekhi Fiz. Nauk* **97**, 453.

Tverskoy, B. A.: 1972, in *Solar Terrestrial Physics / 1970. Part III: The Magnetosphere* (ed. by E. R.

Dyer and J. G. Roederer), Reidel, Dordrecht, p. 297.

Vasyliunas, V. M.: 1968a, *J. Geophys. Res.* **73**, 2529.

Vasyliunas, V. M.: 1968b, *J. Geophys. Res.* **73**, 2839.

Vernov, S. N., Chudakov, A. E., Vakulov, P. V., and Logachev, Yu. I.: 1959, *Dokl. Akad. Nauk, SSSR* **25**, 304.

Vernov, S. N., Gorchakov, E. V., Kuznetsov, S. N., Logachev, Yu. I., Sosnovets, E. N., and Stol-povsky, V. G.: 1969, *Rev. Geophys.* **7**, 257.

THE STRUCTURE OF THE PLASMASPHERE ON THE BASIS
OF INDIRECT MEASUREMENTS

R. A. HELLIWELL

Stanford Electronics Laboratories, Radioscience Laboratory, Stanford, Calif., U.S.A.

Abstract. Using whistler data the large-scale structure of the plasmasphere has been mapped in the equatorial plane. With the aid of topside sounding data the distribution along field lines was found to have a diffusive equilibrium form. The outer boundary of the plasmasphere called the plasmapause, reaches its maximum distance from the earth in the early evening hours and the radial profile shows variations apparently connected with convective motions. Sometimes a well-defined peak in electron density appears near $L = 4$, as a result of convective motion. The properties of small-scale field-aligned irregularities, or ducts, extending between the hemispheres have been measured using topside sounders and whistlers. Latitudinal thicknesses (at low altitude) of these ducts extend from less than a kilometer to tens of kilometers, with spacings up to ten times their thickness. Density variations are 1–10%. From drift motion of whistler ducts, east-west electric fields of magnitude 0.1–1.0 mV/m at $L = 4$ are deduced. During a substorm, plasma is somehow removed from the outer part of the plasmasphere leaving a void called the 'trough'. Replenishment of the trough ionization has been measured using whistlers, and occurs by upward flow from the ionosphere. The time required is in excess of eight days and hence the plasmasphere is seldom in equilibrium with the ionosphere. The downward flux from the plsmasphere at night, measured by whistlers, is sufficient to account for the nighttime F layer.

1. Introduction

The purpose of this paper is to review our knowledge of the structure of the plasmasphere on the basis of indirect methods of measurement. The plasmasphere is the inner, high-density core of the magnetosphere.

The inner boundary of the plasmasphere is arbitrarily taken to be the geocentric spherical shell at 1000 km altitude, as indicated in the sketch of Figure 1. The outer boundary is the field-aligned surface known as the plasmapause, at which the density of electrons drops to a relatively low value. This low density region outside the plasmapause is called the magnetospheric 'trough', and is contained within the region of 'closed' magnetic field lines. Beyond the trough lies the high-latitude region of 'open' field lines where the ionospheric plasma may flow out of the magnetospheric tail.

Identification of the plasmapause as a permanent feature of the magnetosphere was based on 'knee' whistlers recorded as early as 1958 (Carpenter, 1962, 1963). (The term 'knee' refers to the sharp drop in the radial profile of electron density.) At about the same time ion trap data taken on the first Soviet lunar rockets in 1959 were interpreted as showing evidence of a "significant increase of negative gradients of ion concentration in the range of altitudes 15000–20000 km" (Gringauz *et al.*, 1960, 1961).

Further application of the whistler method revealed a characteristic increase in the radial distance to the plasmapause in the evening sector of the magnetosphere (Carpenter, 1966). Study of the time variation of ducted whistlers led to discovery of the radial component of plasma convection driven by perpendicular electric fields (Carpenter and Stone, 1967).

Both the indirect and direct methods of measurement have continued to add to our knowledge of the plasmapause in a complementary fashion. The satellite probe technique has the advantage of frequent sampling in space but cannot give detailed information on the time variation at a fixed location. Ground whistlers can provide data nearly continuously in time at the locations of observable ducts. As a result it is possible to map the plasmasphere in a fairly comprehensive manner using a network of ground stations.

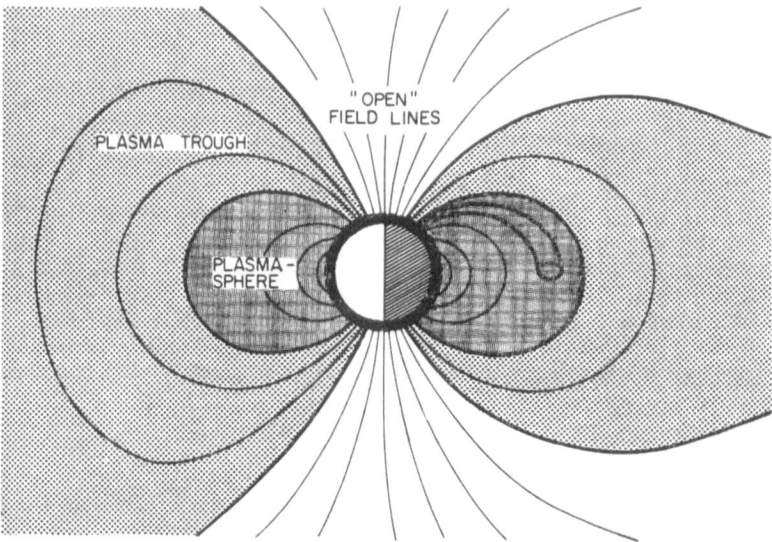

Fig. 1.   Plasmasphere; plasma trough; 'open' field lines; the plasmapause is the field-aligned boundary separating the plasmasphere from the plasma trough.

2. Methods of Measurement

Data on the structure of the plasmasphere can be obtained indirectly by several methods, all based on electromagnetic effects. In general it is easier to find the location of the plasmapause than to measure the density of electrons. Methods of study include Thomson scatter radar (Evans, 1969), high-frequency radio beacon signals from satellites (da Rosa and Garriott, 1969), natural resonances at HF (Calvert and McAfee, 1969) and VLF (Barrington, 1969), micropulsations (Kenney *et al.*, 1968), natural VLF noise (Carpenter *et al.*, 1968), topside sounding (Chan and Colin, 1969), and whistlers (Carpenter and Smith, 1964). With the exception of topside sounding and whistlers, these methods have as yet provided only limited information on the plasmasphere and will not be considered further. Data from topside sounding are of primary importance, but have been limited to the lower part of the plasma-sphere, except in the case of the high-frequency duct phenomenon (Muldrew, 1967).

Much of our present knowledge of the plasmasphere is based on the whistler method, illustrated in Figure 2. Three ducted nose whistlers are shown in part (a). The corresponding field-aligned ducts are shown in (b) and the equatorial electron density profile is shown in (d). Note that the trace labeled '3' shows the smallest delay because of the low density outside the knee. From similar data obtained at other local times, the plasmapause contour of (c) can be constructed.

The reason that whistlers can provide data at great heights when other radio methods (e.g., incoherent scatter and radio beacons) are less effective arises from the effect of the earth's magnetic field. Since the whistler group delay increases as the earth's field decreases a given amount of ionization at high altitude will have a greater effect than the same amount at low altitude (Carpenter and Smith, 1964). In commonly used models of the magnetosphere it is found that 80% of the delay occurs within ± 30 degrees of the equator. The ionosphere itself contributes relatively little to the delay over mid- and high-latitude paths. Thus the observed delays are governed primarily by plasma located near the equator.

The position of the plasmapause can be detected through changes in occurrence of various VLF phenomena. Thus it has been found that as a satellite VLF receiver crosses the plasmapause there is a sudden cutoff of plasmasphere whistlers and an onset of VLF noise (chorus) (Carpenter *et al.*, 1968). Although the exact connection between the gradient of ionization and the changes in the occurrence of whistlers and noise is not yet known, such changes are usually closely associated with the position of the plasmapause as determined from whistler dispersion and direct measurements.

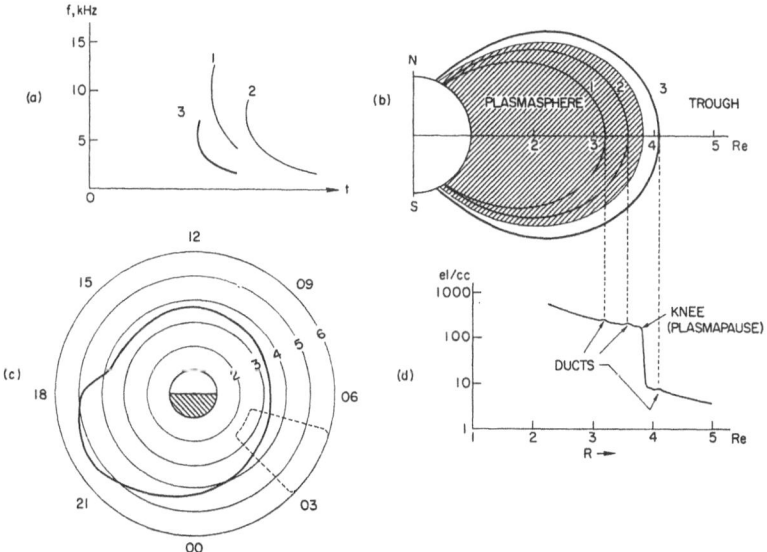

Fig. 2. Plasmasphere mapping using nose whistlers.

3. Large-Scale Structure of the Plasmasphere

By combining topside sounding data (Chan and Colin, 1969) with whistler data (Angerami, 1966), the form of the electron density distribution along a field line can be obtained. An idealized model of the plasmasphere is shown in Figure 3, in which

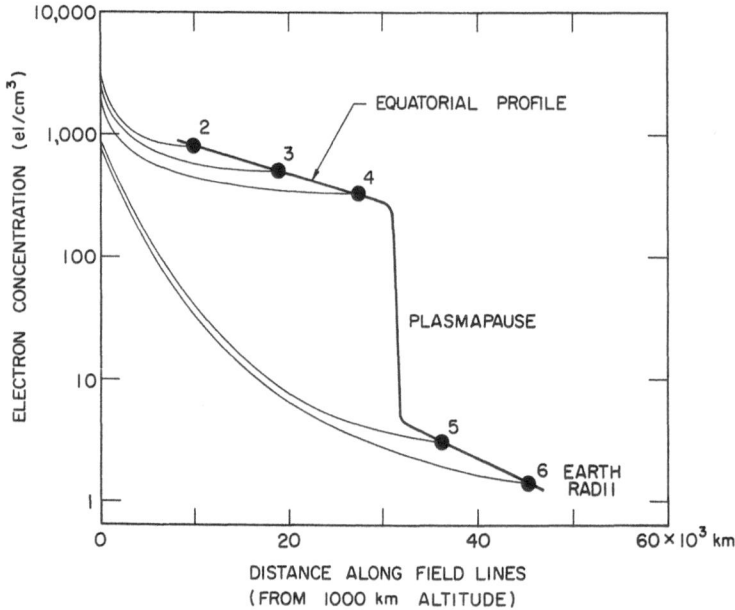

Fig. 3.   Electron density along field lines and in the equatorial plane.

the electron density along field lines is compared with that in the equatorial plane. These curves are not intended to give an accurate representation of electron density, but simply to illustrate the effect of collisions on the distribution of electrons. The computations for Figure 3 were made by assuming an equatorial profile and extrapolating down the field line assuming diffusive equilibrium inside the plasmapause and an R^{-4} variation outside (Park, 1970b). (The ion abundances at 1000 km were 60% O, 1% He and 39% H.) Within the plasmapause the presence of collisions causes the density to fall relatively slowly with distance along the field line. Outside the plasmapause the plasma is in the collisionless state and the density falls much more rapidly, as the curves indicate. It is seen that the equatorial profile depends on the density at 1000 km and on the shape of the distribution along the field lines.

A collisionless distribution of electron density, such as illustrated in Figure 3 for the region beyond the plasmapause, is a relatively temporary occurrence. It is likely to exist only for a short period (less than 1 day) after a substorm. As the magnetospheric void begins to fill through an upward flow of plasma from the ionosphere, the density distribution rapidly acquires a diffusive equilibrium form. Thus even though the plasmapause may be identified for many days during the refilling process, the shapes of the distributions on the two sides of this boundary are the same much of the time.

Equatorial profiles of electron density show considerable variation with local time. The evening bulge of the plasmasphere, illustrated in Figure 4, is perhaps the outstanding regular feature of the plasmasphere. As the earth rotates under this pattern, the equatorial profile as observed from the ground will change as indicated by the three profiles, sketched for times t_1, t_2 and t_3, respectively (Carpenter, 1970). Since the instantaneous whistler viewing area from Eights Station extends over about $\pm 15°$ of longitude (shown by the wedge-shaped areas in the equatorial plane), it is possible to see both profiles simultaneously. Such observations provide further evidence of the spatial nature of the bulge.

In addition to the local time variations connected with the evening bulge, other variations are seen that appear to be connected with convective motions in the plasmasphere. For example Figure 5 shows data obtained at Eights Station in which the equatorial electron density reaches a well-defined peak at $L=4$ (Park and Carpenter, 1970). The peak-to-valley ratio is about $2:1$. An explanation of this peak, or layer,

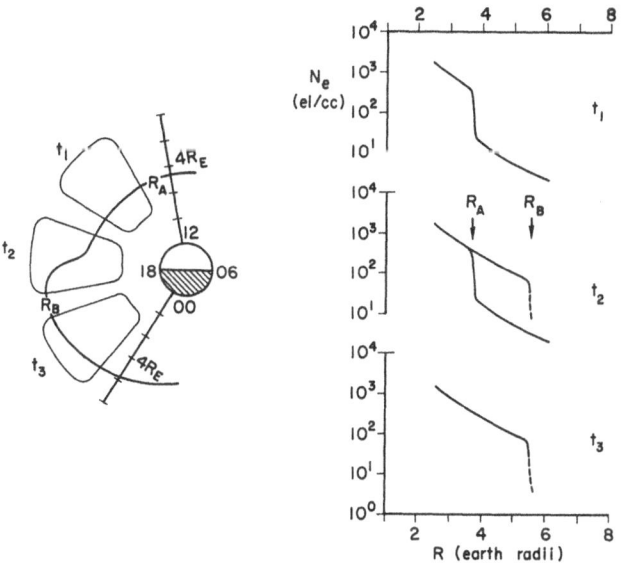

Fig. 4.   Detection of bulge through overlapping knee profiles.

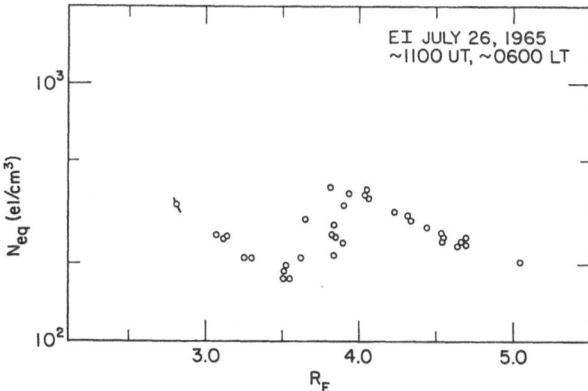

Fig. 5.   Equatorial electron-density profile in the plasmasphere showing a density peak near 4 earth radii geocentric distance. The short bar near 2.8 earth radii represents uncertainty due to measurement error. (After Park and Carpenter, 1970.)

will be outlined later. Other large-scale structures are seen in which the density variations may exceed 3:1 over 30° of longitude.

Because of the very large fluctuations in the plasmasphere associated with substorm activity, it is difficult to detect the average temporal variations. Diurnal variations are thought to be small, of order 30% (Park, 1970a). An annual variation, originally found in low frequency dispersion data (Helliwell, 1961) has been confirmed by nose whistler data (Carpenteı, 1962; Park, 1970a). The density at the December solstice may be as much as twice that in June, depending on the year. An explanation for the annual variation has not yet been found, but the global north-south asymmetry introduced by the relatively heavy concentration of ocean areas at mid-latitudes in the southern hemisphere might conceivably play a role. Thus near the December solstice, the southern hemisphere receives maximum solar illumination, possibly causing increased evaporation of water and hence a larger vertical flux of hydrogen through dissociation.

4. Small-Scale Structure

Small-scale irregularities have also been discovered in the plasmasphere. Conjugate echoes observed with the high frequency topside sounders (Muldrew, 1967) demonstrate the existence of field-aligned irregularities that extend between the hemispheres. Data from Explorer XX (Loftus *et al.*, 1966) on the latitudinal thickness of conjugate ducts are shown in Figure 6 and indicate dimensions of order several km. Duct separations are shown in Figure 7 and tend to be roughly ten times the duct thickness. Percentage variations of density in these irregularities as deduced from the propagation data are of order 1%.

A somewhat larger field-aligned irregularity is found in connection with ducted

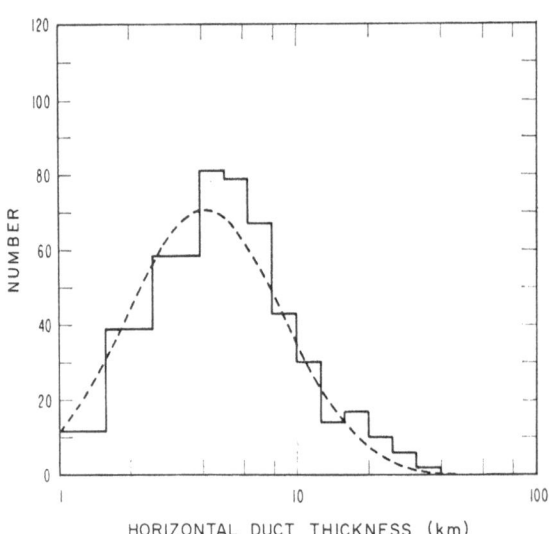

Fig. 6.   Horizontal thickness of HF ducts.

whistler propagation. In Figure 8 are shown the cross sections of ducts in the equatorial plane as deduced from whistlers observed by the OGO 3 satellite (Angerami, 1970). The latitudinal dimensions are several hundred km in the equatorial plane, while the enhancement factors are 5–10%. A connection between HF ducts and the larger whistler ducts has not yet been established. However it is notable that HF northscatter echoes from F-region irregularities have been correlated with the occurrence of man-made whistler-mode echoes (Carpenter and Colin, 1963). These irregularities may be connected with both the HF and whistler field-aligned ducts.

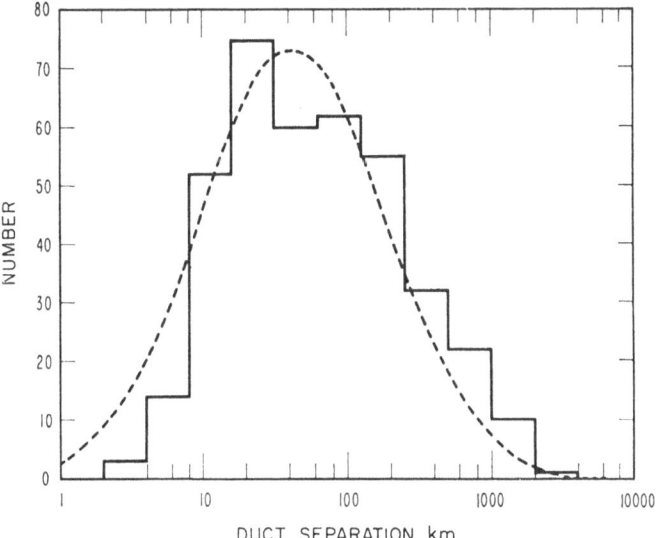

Fig. 7.   Separation of HF ducts.

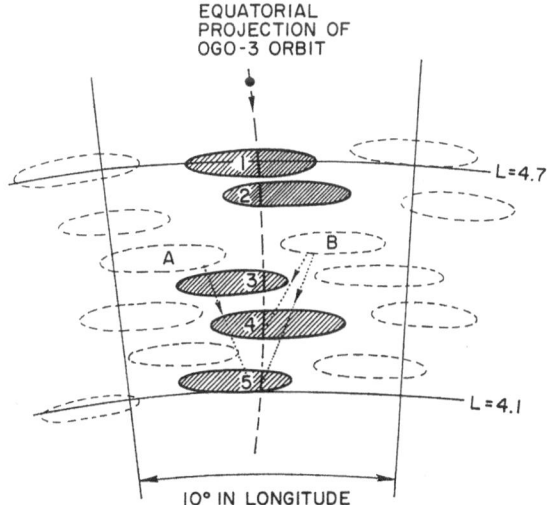

Fig. 8.   Cross sections of whistler ducts in equatorial plane.

5. Dynamic Behavior of the Plasmasphere

Using whistler ducts as tracers, the drift of ionization in the plasmasphere perpendicular to the field lines (cross-L drift) can be measured (Carpenter and Stone, 1967). Latitudinal movements vary from 0.1 R_E/h for relatively quiet times to 1.0 R_E/h for highly disturbed times. East-west electric fields of order 0.1 to 1.0 mV/m on the equator at $L=4$ are deduced from these drift measurements. These fields appear to play an important role in the mixing and erosion of plasma during substorms.

To show how cross-L drift can modify the electron density distribution we consider the idealized curves of density and tube content of Figure 9. Assume that a constant east-west electric field is turned on when the profiles are given by the middle curves. A positive field drives the plasma inwards at a velocity that increases with radial distance (since $v_{drift} = E/B$). Because the ionization cannot easily escape from the tube, the content of the tube tends to remain constant over short periods of time. Since the initial distribution of tube content increases with radial distance in Figure 9, the inward movement causes the equatorial density to develop a peak, as shown. This process is thought to explain the observations shown in Figure 5. The reverse effect occurs when the east-west field is reversed, as shown in Figure 9. Similar drifts in the

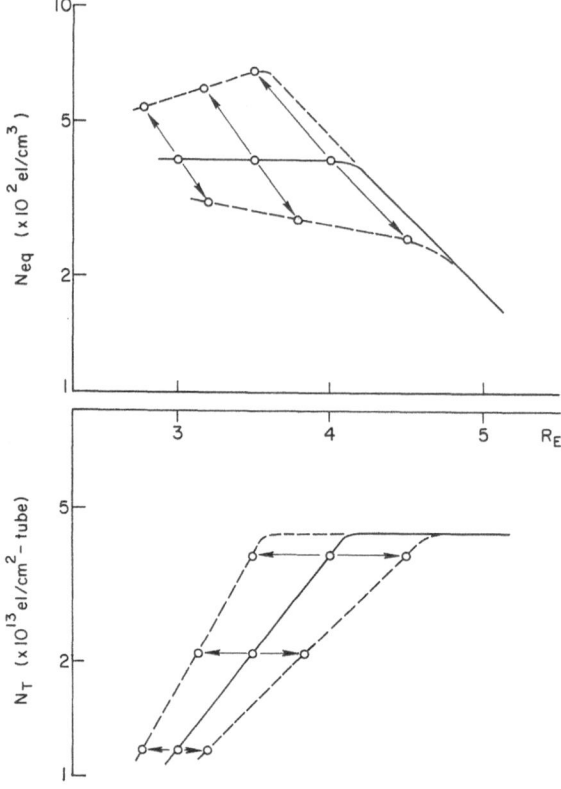

Fig. 9.   Equatorial density and tube content changes caused by constant E field.

longitudinal direction are produced by north-south electric fields, causing complex variations in density.

A further effect of electric fields is seen in the behavior of the evening bulge during development of a substorm (Carpenter, 1970). Figure 10 shows how the bulge tends to rotate in the direction of the earth's rotation when the plasmasphere is recovering after a substorm. During the quieting period the bulge is released and moves to later times. During substorms the bulge is driven towards the sun, against the corotation force. The bulge surges back and forth in response to variations in substorm activity.

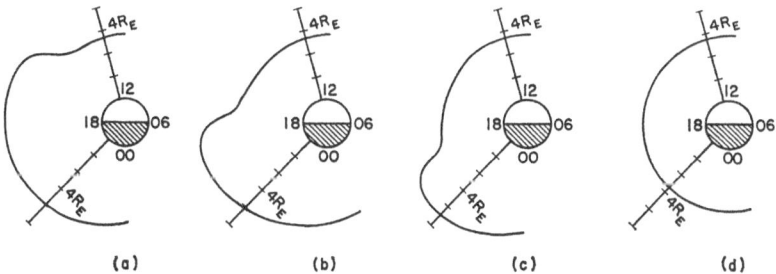

Fig. 10.   Untrapping of bulge during quieting following substorm.

Removal of ionization from the outer region of the plasmasphere appears to take place during a substorm and is the cause of the plasmapause. Although the existence of the plasmapause is strikingly demonstrated by the data, the process by which electrons are removed is much less clear. One possibility is that electrons are removed through transport of ambient plasma from outer to inner parts of the 'noisy' region within the magnetosphere (Cole, 1964). Another is that convection patterns driven by electric fields carry the plasma into the magnetospheric tail region (Axford, 1969). This is a complex question and is the subject of much study at the present time.

Following erosion the plasmasphere slowly recovers through upward flow of plasma from the ionosphere. An example of the recovery phase is shown in Figure 11, which gives the daytime tube content N_T vs L value following the substorm on 15 June 1965 (Park, 1970a). For eight days following the storm there were no disturbances, and it was possible to trace the recovery of the plasmasphere in the region $3.5 < L < 5$. As Figure 11 shows, the recovery rate was approximately the same throughout this range of L. By separating day and night data, the day and night protonospheric fluxes were found. During daytime the upward flux averaged 3×10^8 el/cm² sec while at night the downward flux was about 2×10^8 el/cm² sec. This latter value is sufficient to account for the nighttime F layer. On the eighth day after the main storm there was a small disturbance and the plasmasphere tubes were depleted. However, before that time the density had already reached a value above the monthly median and had not yet leveled off. These results suggest that the recovery time of the plasmasphere following a substorm exceeds eight days. This means that the plasmasphere is seldom, if ever, in equilibrium with the ionosphere.

As we have seen in Figure 11, the plasmapause moves to a higher L value not by

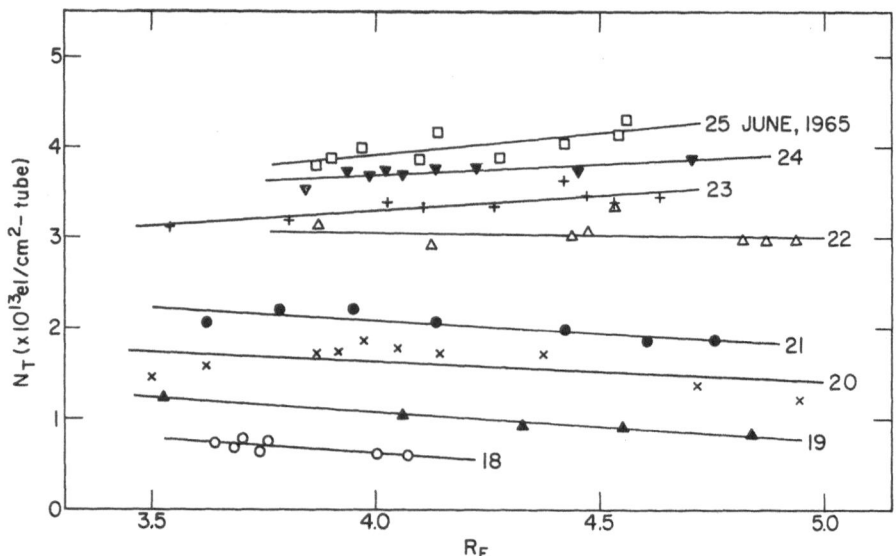

Fig. 11. Nighttime tube content versus tube equatorial radius during the recovery period following the substorm of June 15, 1965. The numbers inside the figure are UT days. (After Park, 1970.)

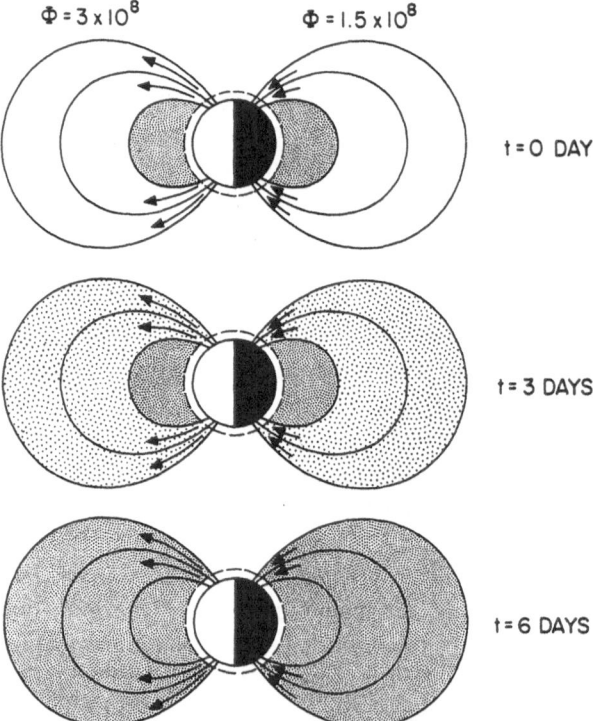

Fig. 12. Sketch of recovery of plasmasphere following a substorm. Ionization flows upward into the plasmasphere from the ionosphere during daytime and downward at night.

expansion but by filling from the ionosphere. During recovery the 'new' and 'old' plasmapauses exist simultaneously giving rise to a double knee in the profile of electron density.

To clarify the points discussed above, the diurnal and day-to-day features in the recovery of plasmasphere after a substorm are sketched in Figure 12. Here the daytime filling and nighttime emptying of the plasmasphere are shown at 0, 3 and 6 d following nearly complete emptying of the plasmasphere beyond $L=2.5$. The diurnal variation of density in the ionosphere is usually large compared with the diurnal variation in the plasmasphere. The result therefore is that the upward and downward fluxes are relatively independent of the plasmasphere density, as suggested by Figure 11.

Acknowledgements

Helpful comments in preparing the paper were provided by D. L. Carpenter and C. Park. The research was supported in part by the National Science Foundation Office of Polar Programs.

References

Angerami, J. J.: 1966, SU-SEL-66-017, Radioscience Lab., Stanford Electronics Labs., Stanford University, Stanford, Calif.
Angerami, J. J.: 1970, *J. Geophys. Res.* **75**, 6115.
Axford, W. I.: 1969, *Rev. Geophys.* **7**, 421.
Barrington, R. E.: 1969, *Proc. IEEE* **57**, 1036.
Calvert, W. and McAfee, J. R.: 1969, *Proc. IEEE* **57**, 1089.
Carpenter, D. L.: 1962, *J. Geophys. Res.* **67**, 3345.
Carpenter, D. L.: 1963, *J. Geophys. Res.* **68**, 1675.
Carpenter, D. L.: 1966, *J. Geophys. Res.* **71**, 693.
Carpenter, D. L.: 1970, *J. Geophys. Res.* **75**, 3837.
Carpenter, D. L. and Smith, R. L.: 1964, *Rev. Geophys.* **2**, 415.
Carpenter, D. L. and Keppler Stone: 1967, *Planetary Space Sci.* **15**, 395.
Carpenter, D. L., Walter, F., Barrington, R. E., and McEwen, D. J.: 1968, *J. Geophys. Res.* **73**, 2929.
Carpenter, G. B. and Colin, L.: 1963, *J. Geophys. Res.* **68**, 5649.
Chan, K. L. and Colin, L.: 1969, *Proc. IEEE* **57**, 990.
Cole, K. D.: 1964, *J. Geophys. Res.* **69**, 3595.
Da Rosa, A. V. and Garriott, O. K.: 1969, *J. Geophys. Res.* **74**, 6386.
Evans, J. V.: 1969, *Proc. IEEE* **57**, 496.
Gringauz, K. I., Bezrukikh, V. V., Ozerov, V. D., and Rybchinskii, R. E.: 1960, *Soviet Phys.–Doklady* **5**, 361.
Gringauz, K. I., Kurt, V. G., Moroz, V. I., and Shklovskii, I. S.: 1961, *Soviet Astron.–AJ* **4**, 680.
Helliwell, R. A.: 1961, *Ann. Geophys.* **17**, 76.
Kenney, J. F., Knaflich, H. B., and Liemohn, H. B.: 1968, *J. Geophys. Res.* **73**, 6737.
Loftus, B. J., Van Zandt, T. E., and Calvert, W.: 1966, *Ann. Geophys.* **22**, 530.
Muldrew, D. B.: 1967, *Can. J. Phys.* **45**, 3935.
Park, C. G.: 1970a, *J. Geophys. Res.* **75**, 4249.
Park, C. G.: 1970b, SU-SEL-70-022, Radioscience Lab., Stanford Electronics Labs., Stanford University, Stanford, Calif.
Park, C. G. and Carpenter, D. L.: 1970, *J. Geophys. Res.* **75**, 3825.

ON THE INTERACTION BETWEEN THE MAGNETOSPHERE
AND THE IONOSPHERE

B. HULTQVIST

Kiruna Geophysical Observatory, S-981 01 Kiruna 1, Sweden

Abstract. Relations between certain structures in the magnetosphere and the ionosphere are discussed in the light of experimental evidence as well as from the point of view of the physical mechanisms thought to cause the structures. New evidence for the fairly frequent occurrence of magnetic-field-aligned electric fields is described. The interpretation of these data in terms of an electrostatic field in or near the upper ionosphere, associated with the temperature difference between the magnetospheric and ionospheric plasmas, is discussed in some detail in the main body of the paper.

1. Introduction

That phenomena in the earth's ionosphere are related to the interaction of energetic particles with the geomagnetic field far away from the earth was an important constituent in Birkeland's thoughts about the aurora already at the end of the previous century. Today's magnetospheric models may be considered as developments of the theory of Chapman and Ferraro (1931, 1932, 1933). The data collected from ground measurements during IGY and afterwards and from more than a decade of satellite and rocket observations have much improved our knowledge of the relations of various ionospheric phenomena as seen from the ground, including aurora and polar magnetic disturbances, to phenomena in space near the earth.

The subject of interactions between magnetosphere and ionosphere is very large indeed. It involves most parts of the magnetospheric and ionospheric physics fields, and cannot be treated in any complete way in one review at this symposium. In fact, many of the most interesting phenomena in this connection have been dealt with here in separate reviews. This is true for e.g. substorms (Akasofu, Feldstein, Dessler) the plasmasphere and plasmapause (Gringauz, Helliwell), and polar wind (Mange) and the F region during magnetically disturbed conditions (Obayashi). In this paper a brief introductory review will first be given of the present knowledge of relations between various spatial structures observed in the ionosphere and magnetosphere and of physical mechanisms thought to be responsible for the existence of these structures. This is intended to provide the background for the main part of the paper, which will deal with an aspect of the interaction between the hot plasma from the plasma sheet in the magnetosphere and the cold ionospheric plasma in high latitudes, an aspect that has not been dealt with in any of the other reviews.

2. Relations between Spatial Structures in the Ionosphere and the Magnetosphere: Experimental Evidence

The 1960's have seen a complete change in our picture of the spatial configuration

Dyer (ed.), Solar-Terrestrial Physics/1970: Part IV, 176–198. All Rights Reserved.
Copyright © 1972 by D. Reidel Publishing Company.

for the plasma around the earth – obtained by means of satellites – and of the morphology for various disturbance phenomena in the high-latitude ionosphere – obtained mainly from ground-based measurements during and after IGY. These new views were summarized by Akasofu (1968) in a schematic diagram shown in Figure 1, where the upper part illustrated the structures seen in the upper atmosphere over the polar cap and the lower part shows the corresponding midnight meridional cross section. This schematic figure is certainly oversimplified, but it still contains some of the main important features in the present views on the relations between certain boundaries (large gradients) observed in the ionosphere and in the magnetosphere. Figure 1 is, however, somewhat controversial in some respects. The observational evidence for the various boundaries and their relations will be briefly described below.

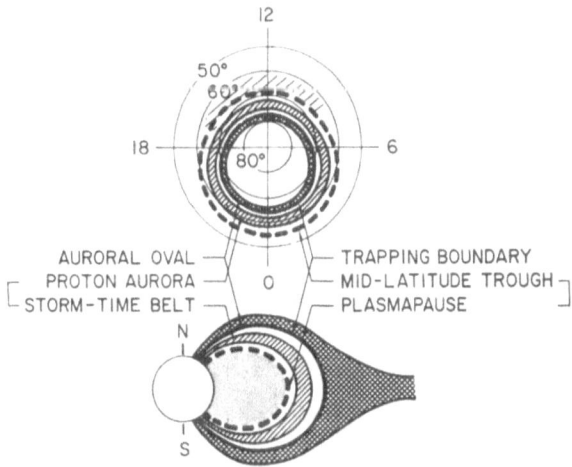

Fig. 1.   Schematic diagram showing the three circumpolar oval structures and their relation to the corresponding magnetospheric structures (after Akasofu, 1968).

2.1. PLASMA SHEET – AURORAL OVAL

The IGY data first offered clear evidence suggesting that the precipitation curve for auroral particles at a given time has an oval shape consisting of two combined spirals, unwinding in opposite directions (see Feldstein, 1960, 1963, 1964, 1966; Feldstein and Solomatina, 1961; Khorosheva, 1961, 1962; Akasofu, 1964, 1965, 1966a, b). These results were based exclusively on observational data obtained from the ground and were presented before the new ideas about the asymmetrical magnetosphere, with their consequences, had become known. It has been clearly shown in recent years that the inner boundary of the plasma sheet in the equatorial plane of the magnetosphere, first observed by Gringauz *et al.* (1960a, b) is connected with the auroral oval in the polar ionosphere by means of geomagnetic field lines. Thus, the plasma from the plasma sheet has been observed by means of satellites practically all the way along the field lines down to the atmosphere (Vasyliunas, 1969). The inner boundary of the plasma sheet connects to the equatorward boundary of the auroral oval. The plasma-sheet boundary appears to be a discontinuity in electron energy but not in density (Vasyliunas, 1968; Schield and Frank, 1969).

There is no observational evidence that there is a cold-plasma component within the plasma sheet (Vasyliunas, 1969).

2.2. Ring current belt – proton aurora belt

The inner boundary of the plasma sheet is a boundary for energetic electrons. The intense proton fluxes characteristic of the plasma sheet reach closer to the earth than the corresponding electron fluxes (Bame *et al.*, 1967; Bame, 1968). At the inner edge of the plasma sheet, where the electron mean energy sharply decreases with decreasing distance, the proton energy instead increases (Frank, 1967a, 1968a, b). The protons may extend almost to the plasmasphere during magnetic storms. They then form the ring current responsible for the main phase of the storm (Frank, 1967b, 1968a), which is called storm-time belt in Figure 1, after Akasofu (1968). The largest proton fluxes occur in the noon-to-dusk sector during disturbed conditions (Frank, 1968b).

The typical energy spectrum of the plasma-sheet electrons has a broad maximum at an energy ranging from a fraction of a keV to more than ten keV (Bame *et al.*, 1967) with the mean energy mostly of the order of 1 keV near the center of the plasma sheet and $\frac{1}{3}$ to $\frac{1}{2}$ of that value near its high-latitude boundaries. The proton spectra have similar shapes, but the mean energy appears to exceed the electron mean energy, sometimes by a factor of up to 10 (Bame, 1968; Hones, 1969).

The angular distributions of both electrons (Vernov *et al.*, 1966; Bame *et al.*, 1967; Hones *et al.*, 1968; Retzler and Simpson, 1969) and protons (Hones, 1968) seem to be very nearly isotropic in the plasma sheet and the ring current belt.

According to Figure 1 the storm-time belt is projected on the atmosphere as a proton-aurora belt, located on the equatorward side of the auroral oval for all local times. There is, however, disagreement on the relative latitudinal location of the proton aurora and the electron aurora in the oval in the midnight and morning parts of the zones. Contrary to what Figure 1 indicates, most experimental results obtained on the ground show that the hydrogen Balmer emissions appear poleward of the discrete auroral forms (Stoffregen and Derblom, 1962; Derblom, 1968; Wiens, 1968). Recent results from 6 keV proton measurements on board the satellites ESRO 1A and ESRO 1B also show the opposite relative location of proton and electron aurora in the evening and the morning sectors (Riedler *et al.*, 1970), in accordance with the results obtained from the ground. The somewhat separated precipitation patterns for electrons and protons of energies above a few keV, with a crossover around midnight, may be a consequence of the convection models for the magnetosphere, if the electric field equivalent to the convection is perpendicular to the sun-earth line (see Axford and Hines, 1961).

For the lowest energies, at which gradient drift is not significant, the distribution in local time might be expected to be that shown in the upper part of Figure 1. That this is not so has been demonstrated at this symposium in the review by Vasyliunas.

2.3. PLASMAPAUSE – TROUGH

If we go further towards the earth in the lower part of Figure 1 we arrive at the plasmapause, the boundary of the plasmasphere. The first clear indication of the existence of this sharp gradient for the plasma density in the inner magnetosphere appears to have come from the Lunik plasma experiments carried out by Gringauz and his colleagues (Gringauz *et al.*, 1960a, b; Gringauz, 1961; Bezrukikh and Gringauz, 1966). Carpenter discovered the plasmapause (the 'whistler knee') independently in whistler data.

Reber and Ellis (1956) from cosmic noise measurements and Warren (1963), Thomas and Sader (1963, 1964) and Dayharsh and Farley (1965) reported significant depletions in the electron density above some subauroral latitude from Alouette topside soundings. Some examples are shown in Figure 2. Muldrew (1965) showed, on the

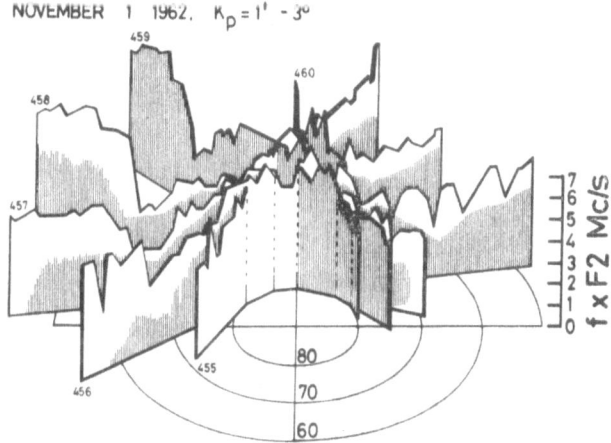

Fig. 2.   Photograph of a 3-dimensional scale model of the quantity $fx\,F2$, the extraordinary ray critical frequency of the F2 layer for a series of consecutive passes of Alouette 1 in November 1962, The quantity $fx\,F2$ can be used to indicate the actual variation of electron concentration NmF2 near the peak of the F2 layer (= 300 km). Circles of geographic latitude are shown (after Thomas and Andrews, 1968).

basis of Alouette data, that 'troughs' in the electron density at the peak of the F layer are a persistent feature of the mid- and high-latitude ionosphere on the nightside of the earth. Liszka (1967) demonstrated that the trough is present in total electron content measurements, i.e. the phenomenon is not simply a vertical redistribution of ionization. Results from the Alouette 2 topside sounder (Nelms and Lockwood, 1967; Hagg, 1967) have shown that at altitudes of 2000–3000 km there is a fall in electron concentration at about the latitude of the F-region trough, but that polewards of this there is usually no increase in the electron density as is found in lower altitudes. Hagg (1967) was able to show that in this low density region the electron concentration was often less than 30 electrons/cm^3 and frequently was as low as 8–15 electrons/cm^3 (Figure 3). Hoffman (1968, 1969a) found that on several occasions when Hagg observed very low electron densities the predominant ion was O^+ even near 3000 km and H^+ ions were streaming upwards. Densities less than 100 electrons/cm³ occurred on about 12% of all ionograms recorded at geomagnetic latitudes above 60° N during the winter periods between December 1966 and February 1968 (Timleck and Nelms, 1969).

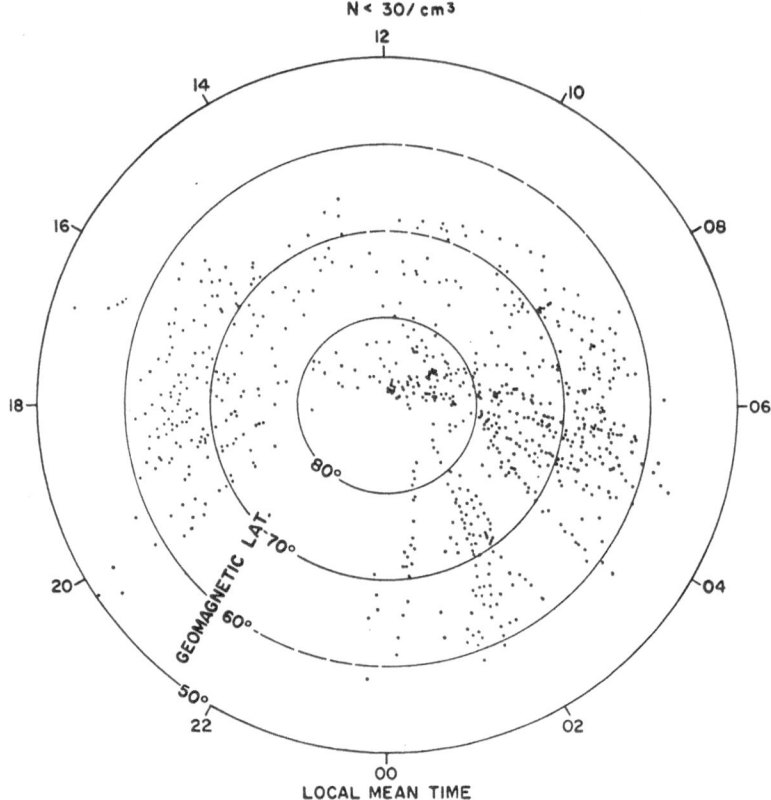

Fig. 3.   Locations (in dipole latitude and time coordinates) where a very low electron density (< 30 electrons/cm³) was observed at 1500–3000 km altitude by Alouette II (After Hagg, 1967).

Muldrew (1965) suggested that the main trough is an extension along the magnetic field lines of the plasmapause. There are, however, some important differences between the behaviour of these two boundaries. Thus, their diurnal variations are quite different (see Figure 4). The increase of the latitude of the projection of the plasmapause on to the polar cap after 1800 h, seen in Figure 4, corresponds to a bulge of the plasmasphere in the evening sector. According to Thomas and Andrews (1968) this difference can be understood in terms of the deformation of the geomagnetic field lines, which the solar wind causes. It, therefore, appears that the geomagnetic field line connection between the plasmapause near the equatorial plane and the subauroral latitude trough in the ionosphere is established.

In conclusion: Whereas the ionosphere inside the plasmasphere is in approximate diffusive equilibrium, experimental data from knee whistlers have been found to correspond to an R^{-4} distribution poleward of the trough (Angerami and Carpenter, 1966). Such a distribution approximates what one expects in a collisionless ionosphere-magnetosphere plasma (cf. Bauer, 1969 and Eviatar *et al.*, 1964). These density distributions fall very much short of the diffusive equilibrium distribution (see Figure 5). This polar cap ionosphere is thought to be produced to a large extent by energetic particle precipitation. The temperature in the upper part of this ionosphere is of the order of a few thousand degrees.

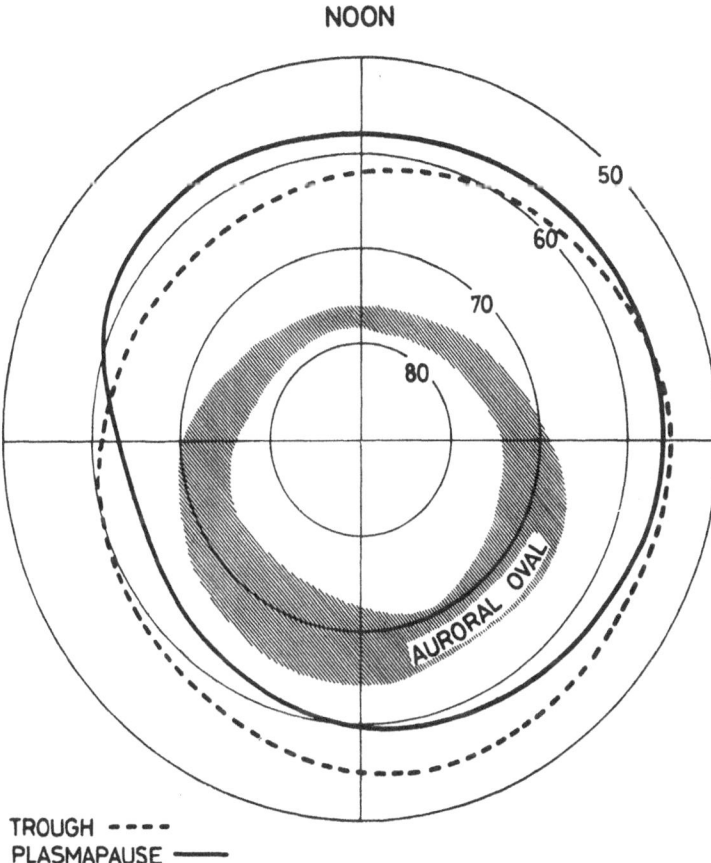

Fig. 4.   The relative positions, in invariant latitude, of the trough boundary (or plasmasphere termination), the auroral oval, and the plasma pause projected along a field line to the ionosphere. The positions given are those expected for moderate magnetic activity ($K_p = 3$).
(After Thomas and Andrews, 1968.)

3. Outline of the Physical Mechanisms Thought to be the Causes of the Boundaries

Already in the original work of Axford and Hines (1961) on the effects of the convective motions in the magnetosphere it was noted that in any reasonable model there

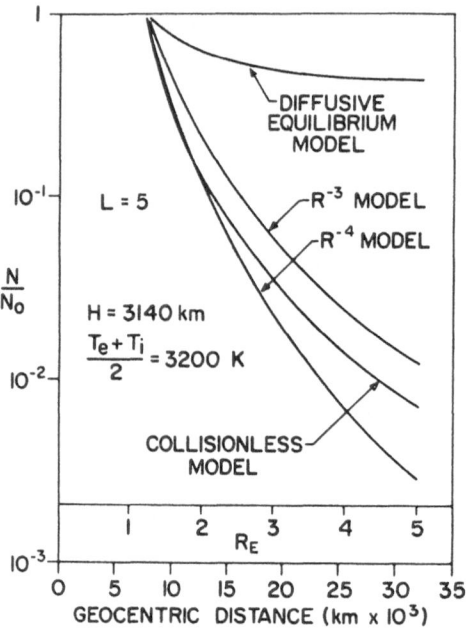

Fig. 5.   Plasma distributions along a magnetic field line whose apex in the equatorial plane is at
$L = 5R_e$ are shown for a diffusive equilibrium, and a collisionless model, as well as for R^{-3} and R^{-4}
power law models (after Bauer, 1969).

would be an inner region of more or less co-rotational motion, outside of which the
plasma convection is the dominating process. Axford and Hines also found that the
co-rotating region would have a bulge on the evening side of the magnetosphere as
was later found to be the case for the plasmasphere (Carpenter, 1966). In their
1961 paper they identified the boundary between the convective and co-rotating
regions with the low latitude edge of the auroral region, i.e. with the inner edge of
the plasma sheet in the equatorial plane.

Carpenter (1962) explicitly proposed that the plasmapause is the boundary of
convection near the equatorial plane instead of the inner plasma-sheet boundary,
and suggested that heating of the outer magnetosphere during magnetic storms might
produce the observed density discontinuity. His idea was that the plasmapause is
connected with the equatorward edge of the auroral belt by the geomagnetic field
lines. However, the plasmapause is sometimes seen as close to the earth in the equatorial plane as at $L = 2.5$, which is certainly too close for the inner edge of the auroral
belt. A resolution of this difficulty was found by Petschek and Kennel (1966) (see
also Kennel, 1969). They demonstrated that it is possible to have an inner boundary
of the electron flux (plasma sheet) which does not correspond to the inner boundary
of the convection.

The convecting field tubes penetrating through the auroral discontinuity to the
plasmapause must evidently be emptied of its plasma-sheet electrons at the plasma-
sheet boundary. Since these electrons are known to be precipitated in auroral latitudes

it is reasonable to assume that precipitation removes the electrons. Petschek and Kennel (1966) proposed that so called strong diffusion into the loss cone creates a boundary to the plasma-sheet electrons at the location where the time scale for the flow and the life time of the electrons become comparable. The work of Petschek and Kennel has been extended and generalized by Vasyliunas (see Kennel, 1969).

If during the flow of a field tube toward the earth the lifetime of the electrons is short compared to the time scale of the flow, many particles will be precipitated and the tube will flow onward depleted of the precipitated particles. Kennel (1969) has estimated the flow time, T_F, and the lifetime for strong pitchangle diffusion, T_M, for a highly simplified situation, which, however, is likely to contain the most important features of the real case. He found that

$$T_F = 10^7/L^2\phi \qquad \text{sec,} \tag{1}$$

$$T_M^+ \approx \left|\frac{m_i}{m_e}\right|^{1/2} \frac{L^4}{E_i^{1/2}} \qquad \text{sec for ions} \tag{2}$$

and

$$T_M^- \approx \frac{L^4}{E_e^{1/2}} \qquad \text{sec for electrons} \tag{3}$$

where L is the McIlwain L parameter; ϕ is the electric potential drop over the magnetosphere corresponding to the convection, in kV; $E_{i,e}$ is the energy of ions and electrons, respectively, in keV and m_i, m_e are the ion and electron masses.

Since $T_F \propto L^{-2}$ and $T_M^\pm \propto L^4$ one expects $T_M^\pm > T_F$ for sufficiently large L. Thus, at great distances precipitation will not affect the density in the field tube significantly. The point at which $T_M^\pm \approx T_F$ determines the boundary, since, beyond this point, particles are depleted very effectively by precipitation before the flow can travel very far.

Some results of slightly more elaborate calculations of T_F and $T_M^\pm$ are shown in Figure 6 (after Kennel, 1969). T_F for three different potential differences across the magnetosphere and T_M for electrons and protons of three different energies are shown. The intersection of T_M (solid) and T_F (broken) curves determines a precipitation boundary for all particles with energies so small that the $\mathbf{E} \times \mathbf{B}$ drift dominates. It may be seen from Figure 6 that

– the low-energy electron boundary lies beyond the plasmapause;
– the boundaries move inward with increasing potential difference across the magnetosphere and presumably, therefore, with increasing magnetic activity;
– of those particles with sufficiently small energies to be carried by the convective flow the higher energy particles precipitate out first, i.e. at slightly greater latitudes;
– the proton boundary is located well inside the electron boundary.

All these conclusions seem to be verified by experimental data, except the third one which is in contradiction with the observations of a regular softening of the energy spectrum of auroral electrons with increasing latitude (Burch, 1968; Maehlum, 1968;

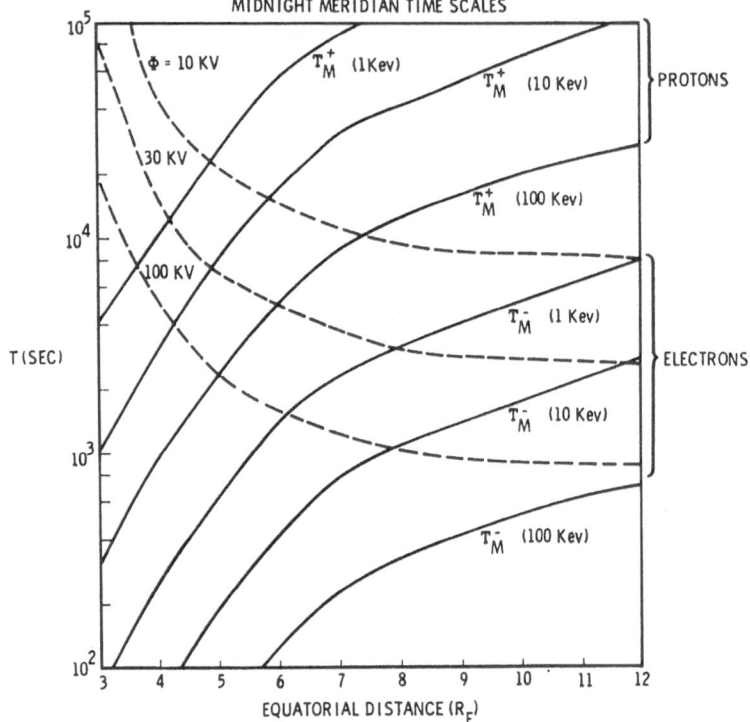

Fig. 6.   Midnight meridian time scales. Shown here are results of crude calculations of several time scales pertinent to uniform radially inward flow along the midnight meridian. The B field was taken dipolarly at small distances and converged to a constant 30γ magnitude at large distances. The dotted lines indicate flow times for various uniform east-west electric fields across the magnetosphere, denoted by total potential differences of 10, 30 and 100 kV. The solid lines indicate minimum lifetimes $T_M\pm$ for protons (+) and electrons (−) at the energies indicated. The intersection of a solid and a dotted line determines the precipitation boundary for particles of a given energy (after Kennel, 1969).

Hoffman, 1969c; Riedler *et al.*, 1970). There is, however more or less regularly a soft equatorward edge region of the precipitation zone (Riedler *et al.*, 1970) which may be related to Kennel's mechanism. The softening on the poleward side must then be caused by some other physical process.

In conclusion it may be said that good evidence has been presented by Kennel and others for strong pitchangle diffusion being the cause of the gap between the inner edge of the plasma sheet and the plasmapause. An alternative proposal has, however, been presented at this symposium by Karlsson (1970). He has carried out calculations for two distinct particle populations, one hot and one cold and has found it possible to describe the existence and locations of both the inner boundary of the plasma sheet and of the plasmapause in terms of drift motions of these particle populations.

Nishida (1966) and Brice (1967) have brought all the above-mentioned ideas together in their models of the magnetosphere. Brice (1967) put emphasis on the plasma inflow to the polar cap ionosphere, whereas Nishida (1966) discussed the loss of plasma along open field lines. He and others suggested that an evaporative mechanism, similar to that which causes the loss of hydrogen atoms from the upper atmosphere, might be involved. Banks and Holzer (1968, 1969a, b, c) and Axford (1968) have shown that something more than ordinary evaporation is involved. The polar wind, as they have called the outflow, depends upon bulk properties of the medium, like electric fields, pressure gradients etc., instead of on details of the distribution function for the gas, as the evaporation does.

The only direct observations of the polar wind made hitherto are those of Hoffman (1968). The earlier mentioned peculiar characteristics of the high latitude ionosphere may, however, be taken as good indirect evidence for the polar wind being an important process, responsible for some of the major differences between the ionosphere within the plasmasphere, which is in diffusive equilibrium with the production of ionization by solar XUV radiation in fairly low altitudes, and the very thin, collision-free upper ionosphere over the polar caps, produced to a large extent by energetic-particle precipitation. As will be briefly discussed below, electric fields may, however, sometimes prohibit the polar wind flow. The effect on the density of such fields is similar to that of the polar wind, so it is at present not possible to draw any conclusions concerning the relative importance of the two effects on the basis of electron-density measurements in the upper ionosphere.

4. Electrostatic Fields along the Geomagnetic Field Lines; Summary of Earlier Work

The controversy about electric fields directed along the geomagnetic field lines is an old one. It started already in the early forties, if not earlier, after Alfvén (1939, 1940) had published his model for the production of aurora and magnetic storms, where electric fields along the magnetic field lines played an important role. The main argument raised against the existence of electric fields directed along the magnetic field lines has subsequently been that the electric conductivity along the field lines is so high that no electric fields can be upheld. That argument has been met by Alfvén and fellow workers with the counter argument that the concept of conductivity is without meaning in a collisionless plasma (see e.g. Alfvén and Fälthammar, 1963, p. 162). They have demonstrated theoretically that under certain conditions electric fields along the magnetic lines of force can be maintained in a collisionless plasma (Alfvén and Fälthammar, 1963, pp. 163–166; Persson, 1963).

It also follows from the discussion of the strong diffusion mechanism of Kennel and Petschek (1966; see also Kennel, 1969) in the previous section that that mechanism leads to different precipitation rates for electrons and ions and, therefore, to a net deposit of charge in the atmosphere. This sets up a potential difference between the equatorial plane and the ionosphere. If the ionosphere could not emit any current this potential difference would grow until the rate of precipitation of negatively- and positively-charged particles equalized and thus no more current would flow along the magnetic lines of force. But then the electrons and protons would be precipitated in the same region and their precipitation boundaries would coincide. The differences in latitudinal distributions for energetic electrons and protons, which have been observed, imply that cold electrons are involved in the keeping up of charge neutrality. The cold electrons seen on the earthward side of the plasmasheet boundary, the number density of which is about the same as the number density of energetic electrons in the plasma sheet (Vasyliunas, 1969), are probably extracted from the ionosphere.

More and more evidence has been presented to prove the importance of field aligned currents in the three-dimensional magnetosphere-ionosphere current systems, as proposed already by Birkeland (1908, 1913). See e.g. Boström (1968) or Akasofu (1968) for reviews.

Although there is thus a lot of indirect evidence in favour of the existence of field-aligned electric fields, only one direct measurement of such a field has hitherto been reported (Mozer and Bruston, 1966). The result of this experiment was, however, rather extreme. This has, in combination with the fundamental difficulties associated with electric-field measurements in the ionospheric plasma, raised some doubts about the reliability of the result, particularly after barium cloud experiments had put quite a low limit on the magnetic-field-aligned electric field component (in a situation free from aurora and associated activities, however; Mende, 1968). It is, therefore, of interest that satellite particle measurements have recently provided new evidence for the existence of field-aligned electric fields quite frequently in auroral latitudes. These results will be briefly described next.

5. Production of a Potential-Drop in or near the Upper Ionosphere over the Polar Caps: Some General Considerations

As has been mentioned above, the plasma in the polar cap ionosphere has a temperature of a few thousand degrees at distances of the order of 1000 km from the earth's surface. In the plasma sheet, on the other hand, the plasma is very hot (see earlier section). The average of the mean electron energy is of the order of 1 keV, corresponding to an equivalent temperature of about 10^7 K. Ten times higher mean energy has also been reported. This hot plasma is in direct contact along the geomagnetic field lines with the cold ionospheric plasma in auroral latitudes. It is well known from laboratory plasma that when two plasmas of different characteristics are in contact with each other a potential difference is produced over a narrow layer in the contact region between the two plasmas, which is sometimes called a double layer. The thickness of such a double layer is of the order of some Debye lengths. In fact, Alfvén proposed already in 1958 the existence of a potential double layer in or just outside the ionosphere. This was well before the discovery of the plasma sheet. Here we will consider the stationary state potential difference between the hot magnetospheric and the cold ionospheric plasmas. It seems unlikely that such an electric field is limited to a layer of only some Debye lengths' thickness, so we will not use the term double layer but will instead call it transition layer.

Let us consider a few quantitative relations.

A temperature gradient in a plasma is equivalent to an electric field due to thermal diffusion

$$\mathbf{E}_{\nabla T} = \frac{\alpha k}{2e} \nabla T \tag{4}$$

(see e.g. Finkelnburg and Maecker, 1956, p. 327), where α is the dimensionless thermodiffusion factor which has a value of 1.4 in a plasma with singly charged ions, k is Boltzmann's constant, e is the electronic charge and T is the temperature. This electric field is produced by the difference in diffusion velocities of the electrons from the different temperature regions. The temperature gradient associated with a pressure gradient gives rise to an electric field which has the value given by (4) if $\alpha/2$ is replaced by 1 in the extreme case of zero density gradient. The potential difference will exist as long as the temperature gradient is upheld.

If we apply (4) to the magnetosphere-ionosphere plasma, assuming the magnetospheric plasma to be quasithermal with an equivalent temperature of the order of 10^7 K, an ionospheric temperature of the order of 10^3 K, and taking $\Delta V = E \cdot d \propto \nabla T \cdot d = \Delta T$ (d is the layer thickness), we find that this average temperature for the plasma sheet will cause a potential drop of 600 V near the ionosphere. The electric field will be directed downwards. If the equivalent temperature of the magnetospheric plasma is instead 10^8 K, corresponding to an average particle energy of the order of 10 keV, which is sometimes seen (see e.g. Bame *et al.*, 1967), the potential drop in the transition layer would be 6 kV. The equivalent 'temperatures' expected in the plasma sheet are not very much higher, so one does not, on the basis of (4), expect the potential drop to exceed 6 kV very often.

It should be remembered that (4) is valid for plasma with collisions, where the concepts of diffusion and conductivity have a meaning. The plasma in the plasma sheet is a collisionless plasma (i.e. the collision mean free path is much larger than the dimensions of the magnetosphere) if one considers ordinary particle encounters. It is likely that field irregularities or waves between the plasma sheet and the ionosphere are efficient enough to replace the collisions in creating a diffusion process in the plasma-sheet horns to the atmosphere. For a thermal diffusion effect to be about the same as in the Coulomb-scattering case the energy dependence of the scattering cross-section has to be about the same. However, even if there is no diffusion, a potential difference may still exist between the plasma sheet and the ionosphere. The fast electrons from the hot plasma may deposit a net negative charge in the ionosphere. The potential difference produced by this net charge will decrease the electron flux and increase the ion flux into the ionosphere and it will draw charges out of the ionosphere in a way similar to what happens in a plasma where diffusion is important. There may be differences between the two cases in voltage drop, thickness of the layer and other parameters. but the basic production mechanism for the potential difference is expected to be essentially the same. The value of the potential drop should depend on the detailed shape of the energy distribution in the plasma sheet.

Although it is, thus, fairly clear that a potential drop along the geomagnetic field lines may be produced initially when a hot magnetospheric plasma is brought into contact with the ionosphere in auroral latitudes, it is perhaps not quite clear that such a potential difference can be upheld in a stationary state. A necessary condition is that a sufficiently high temperature difference between the plasma sheet and the ionosphere should be maintained. This conclusion is applicable also to the diffusion-less case and to cases where we cannot speak about temperature but have to talk in terms of energy distribution. Since the hot plasma is continuously cooled at the ionospheric end, a continuous energy input to the plasma sheet is needed. There are evidently energy sources in the magnetospheric tail, and it seems quite probable that a stationary state sometimes exists.

6. Strong Field-Aligned Anisotropy Observed in Low-Energy Auroral Positive Ions

Some new evidence will now be presented, which clearly indicates that a potential drop of 6 keV or more does indeed occur fairly frequently in auroral latitudes and may have a very long lifetime. The electric field accelerates the local thermal positive ions downwards and produces field-aligned fluxes. The existence of such field-aligned fluxes was first reported last year by Reme (1969) on the basis of a sounding rocket measurement and by Riedler (1969) from measurements on board the satellite ESRO 1A.

Observations of field-aligned ion fluxes have been reported at this symposium by Reme and by Galperin. I will now describe briefly the ESRO 1 observations.

Kiruna Geophysical Observatory had a low-energy auroral particle experiment, consisting of 10 curved plate analyzers with channel multipliers, on each of the satellites ESRO 1A (Aurora) and ESRO 1B (Boreas). These satellites, which were magnetically stabilized, were launched into polar orbits on 3 October 1968 and 1 October 1969, respectively. The apogee was 1533 km and perigee 258 km shortly after launch of ESRO 1A and 378 km and 281 km, respectively for ESRO 1B. The 10 detectors were collected in two groups, measuring at 10° and 80° from the magnetic field lines. Of the 10 detectors 3 on ESRO 1A and 4 on ESRO 1B measured positive ions of about 6 keV energy in directions 10° and 80° and of 1.4 keV energy at 80° from the magnetic field lines on ESRO 1A. On ESRO 1B there was also a detector for 1.3 keV ions in direction 10°. The width of the energy windows of the detectors was about 10% of the nominal energy value and the field of view was equal to or less than ±8 degrees. All ion channels were sampled simultaneously. There were also in each experiment 2 reference detectors with Ni63 radioactive sources in front of them.

On the basis of high speed telemetry data from 202 transits through the auroral particle precipitation zone, of which about 80% were in the evening and midnight sectors (eccentric dipole time) and the remaining ones in the morning-noon sector, Hultqvist et al. (1970) found a number of different types of pitchangle distributions for the positive ions. These are illustrated by Figure 7. In the figure the positive ions have been named protons on the basis of the assumption that protons constitute the majority of them. In fact, the electrostatic analyzer used cannot separate the various ions with the same energy, so the composition of the positive ions is not known.

The depleted-loss cone type of anisotropy was found to occur in all cases from the nightside at the equatorward edge of the particle precipitation zone. Polewards this region was found to be followed by a – mostly fairly narrow – latitude range of isotropy. Further polewards the proton flux inside the loss cone was often found to be much higher than outside it. It is this field-aligned anisotropy that is of special interest here and the observations of it will first be discussed in some more detail.

Figure 7 illustrates a fairly frequent observation on the nightside: an anisotropy

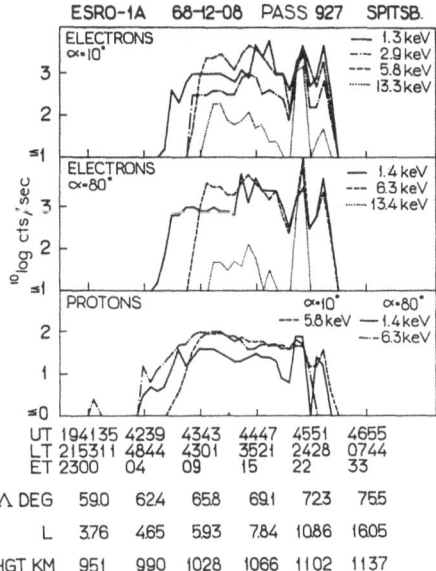

Fig. 7.   Measurements on board the satellite ESRO 1A of electron and positive ion fluxes at various low energies and for two different pitchangles, $\alpha = 10°$ and $80°$. At 1946 UT a field aligned anisotropy in the proton flux is recorded. A curve connects 8 second averages for the count rate of a detector. For the eccentric dipole time (ET) only hours and minutes are given. Λ is the invariant latitude at the earth's surface derived from the L value for the satellite location. The energy windows of the detectors have a width of about 10% of the energy value. The field of view was equal to or less than $\pm 8°$. The telemetry station which received the data is given at the top right. (After Hultqvist *et al.*, 1970.)

with peak flux along the magnetic field lines occurs at the poleward limit of the auroral particle zone. The curves connect points which are averages over 10 subframes of the telemetry format, i.e. over 8 sec of time. The measurements have a temporal resolution of down to 50 msec for fluxes above certain limits. The reason for the choice of the 8 sec average as a basis for the diagram in Figure 7 is – besides the practical one – that plottings with a higher time resolution have been found to provide very little additional interesting information.

The anisotropy in Figure 7 develops mainly by the disappearance of the positive-ion flux in direction 80°. The ion anisotropy coincides in space with a burst of electrons. It should be mentioned that the 10° curve for 6 keV ions is above the 80° curve by a factor of 1.5–2 when isotropy exists, due to differences in conversion factors.

A pronounced positive-ion anisotropy in a poleward boundary electron-flux peak has been seen in about 10% of the 202 cases investigated. Field-aligned anisotropy for the positive ions associated with a region of high electron flux is also seen sometimes inside the main body of the latitudinal distribution. But it does not occur always when there is a strong electron burst, neither at the poleward boundary, nor inside the particle zone. Figure 7 is interesting as it shows that in one and the same region, separated by only some 20 seconds of satellite travel time, there may be an electron spike associated with a strong magnetic field-aligned ion anisotropy and another, even stronger, electron burst which is not associated with any change at all in the positive ions.

The field-aligned positive-ion anisotropy is often due to the flux measured at 80° being lower in a region (of space or time) than in neighbouring regions, rather than the 10° flux being higher as mentioned earlier. This is illustrated still better by Figure 8

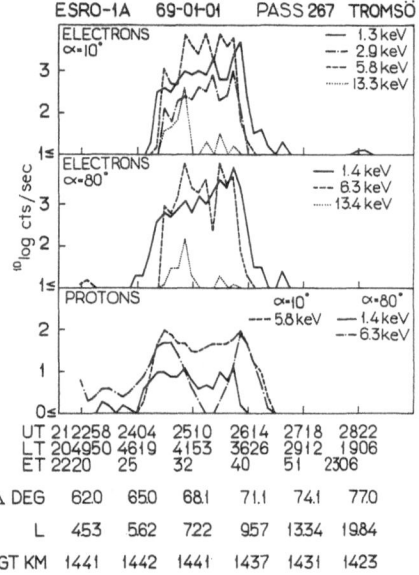

Fig. 8.   In the middle of the precipitation zone there is a strong field-aligned anisotropy in the positive ion fluxes. See caption of Figure 7 for explanation of notations. (After Hultqvist *et al.*, 1970.)

than by Figure 7. As can be seen, in the central region of the particle zone the 80° ion flux is lower than the 10° flux by a factor of more than 40. Except for this low value of the 80° positive-ion flux there are no remarkable features of the latitude distributions for the various electron and ion fluxes shown in Figure 8. On both sides of the region of field-aligned ion anisotropy there are regions with approximate isotropy. In extreme cases the latitude distributions for the 10° and 80° fluxes of 6 keV positive ions appear to be almost unrelated. This is shown in Figure 9. Even there, however, there is a depleted loss cone type of anisotropy on the equatorward side, followed by a narrow region of approximate isotropy, to the north of which the 6 keV ion flux is higher in the loss cone than outside it. This last-mentioned region coincides with that part of the electron zone where the fluxes are highest.

In a six-hour period between 4 p.m. and 10 p.m. UT on 30 November 1968 a strongly anisotropic 6 keV ion flux, with the flux inside the loss cone (at 10° from the field line) being more than 10 times greater than outside it (at 80° pitchangle), covered more or less the entire polar cap. This was seen in four consecutive orbits of ESRO 1A.

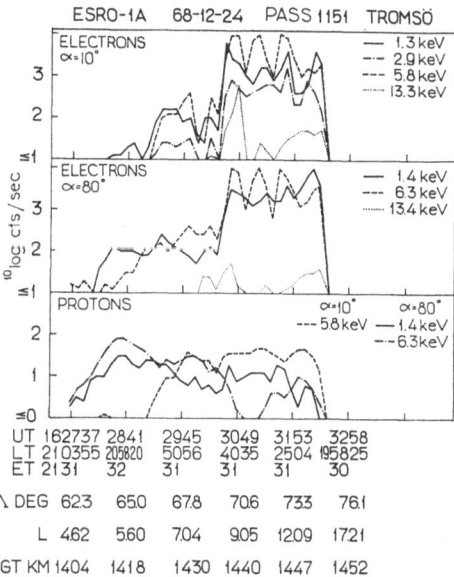

Fig. 9.   A case where the latitudinal distributions for 6 keV positive ion fluxes inside and outside the loss cone are extremely different. See caption of Figure 7 for explanation of notations. (After Hultqvist *et al.*, 1970.)

The measurements taken in one of these orbits are shown in Figure 10. Fairly strong electron fluxes in this period over the whole polar cap were also observed, but the 80° ion flux was always below the sensitivity threshold there. The electron spectrum was fairly hard. In satellite data for half a year or more this extensive coverage of the polar cap by count rates above 10 per second – corresponding to fluxes of the order of 10^5 per cm^2 sec sr keV – was only observed in this event. It is interesting to note that the geomagnetic disturbance level was extremely low in this period. This may perhaps be related to the long known anticorrelation between the occurrence of polar cap aurora and the disturbance level.

We may summarize the results of Hultqvist *et al.* (1970) concerning the field-aligned anisotropy in the evening-midnight and morning-noon sectors as follows:

– It occurs fairly frequently for 6 keV protons. Ratios of the flux at 10° pitchangle to that at 80° of above 10 were seen in 16% of 202 passes.

– It may cover large areas and sometimes be present for hours.

– It occurs on both the dayside and the nightside of the earth.

– It occurs in the same regions as strong auroral electron fluxes.

At energies above 10 keV proton pitchangle measurements have been made by
Sharp *et al.* (1967) with a satellite and Whalen *et al.* (1967) with a rocket. They found
isotropy within a factor of two. Indications of field-aligned proton fluxes have been
obtained from several measurements of the height profile of proton aurora (Johansen
and Omholt, 1963; Miller and Shepherd, 1968; Wax and Bernstein, 1970). Rocket
measurements have thus shown peaks at 120–140 km and the height extension of the
Hβ emission to be only some 20–30 km.

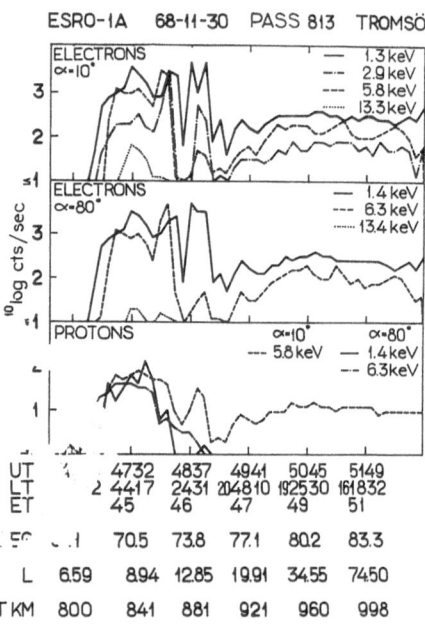

Fig. 10. In a six-hour period on 30 November 1968 ESRO 1A observed energetic positive ions
over most of the polar cap during four consecutive passages. The 6 keV ion flux in the loss cone was
much greater than at about 80° pitchangle for the whole period. See caption of Figure 7 for explanation of notations. (After Hultqvist *et al.*, 1970.)

The reader is referred to Hultqvist *et al.* (1970) for a detailed discussion of the
measurements, and, in particular, for the arguments on the basis of which they have
ruled out the possibility of the effect not being real but due to some instrumental
effects. Here it will only be mentioned that in spite of very careful error analysis no
signs of any erratic functioning of the experiment have been found.

It may be mentioned that the study discussed above was based on only a small
fraction of the total amount of data which will eventually become available. It is,
therefore, to be considered as preliminary and will be followed by a more extensive
investigation based on the entire amount of data and covering all (or at least most)
local times. But even so, some important conclusions about physical mechanisms
may already be drawn on the basis of the 202 passes investigated hitherto. The
following are more or less obvious conclusions:

– The forces accelerating the ions act approximately along the geomagnetic field lines.

– The acceleration takes place fairly close to or in the upper ionosphere.

– The mechanism accelerating the positive ions along the magnetic field lines is also associated with the existence or production of energetic electrons in the field tubes.

The second conclusion is based on the loss cone being only a few degrees wide in the equatorial plane, and that therefore, an extremely good alignment of the accelerating force with the magnetic field lines and with a large increase in the 'parallel' energy for all protons in the field tube due to that force are needed for the production of the observed ratios between $10°$ and $80°$ fluxes, if the mechanism is localized to the equatorial plane. From the relation obtained on the basis of the assumption of the first adiabatic invariant being constant:

$$\frac{\sin^2 \alpha_2}{\sin^2 \alpha_1} = \frac{B_2}{B_1} \cdot \frac{E_1}{E_2} \tag{5}$$

(Catchpoole, 1969), where α is the pitchangle, B is the magnetic field intensity, E is the energy, and 1 and 2 refer to the equatorial plane and the ionosphere, respectively, we can see that a particle with an original pitchangle of $90°$ in the equatorial plane, which arrives in the auroral zone ionosphere $(B_1/B_2 \approx \frac{1}{500})$ with a pitchangle of $10°$ and an energy of 6 keV, has to have an original energy of only 0.36 eV. The perpendicular temperature of the plasma in the equatorial plane thus has to be very low if the acceleration takes place at the equator for the high degree of field-aligned anisotropy to be seen in the ionosphere. We know that the equivalent temperature for the hot plasma in the plasma sheet is of the order of keV, and can therefore draw the conclusion stated above.

The only physical mechanism that seems able to meet all the requirements resulting from the measurements described is acceleration in an electrostatic field directed along the geomagnetic field lines and located in or near the upper ionosphere. One candidate is an electric field associated with the potential difference between the hot magnetospheric plasma and the cold ionosphere discussed earlier. That discussion will now be confirmed.

7. Discussion of some Properties and Consequences of an Electric Field Caused by the Equivalent Temperature Difference between the Magnetospheric and Ionospheric Plasmas

The question of what happens to a plasma in an electrostatic field has been considered by a number of authors (Buneman, 1959; Field and Fried, 1964; Swift, 1965; Block and Fälthammar, 1968; Ossakov, 1968; Stenflo, 1968, 1969a, b, 1970). Different instabilities may be produced in the plasma by the electric field, which may inhibit the free flow of current and create an effective resistivity also in a collisionless plasma. If this occurs in or near the upper ionosphere it will, of course, be much 'easier' for the magnetospheric plasma to maintain the electrostatic field. But even if no such anomalous resistivity is produced a situation compatible with the observations described earlier may exist. In such a case the thermal plasma inside the transition layer will more or less disappear provided the electric field is maintained. A stationary state will finally be achieved when the current drawn by the electric field inside the layer is equal to the diffusion rate through the boundary surfaces of the layer of electrons from below and positive ions from above. The low energy positive ions diffusing in from above will be accelerated by the entire potential difference and will be seen as monoenergetic ions directed along the magnetic field lines in the ionosphere. A 'hot' ion population, originating in the plasma sheet, will, of course, be superimposed on this monoenergetic and monodirectional one.

There does not seem to be any problem involved in having enough low-energy protons available, by diffusion through the upper boundary surface of the layer, to account for the observed proton fluxes. If we assume the number density of thermal protons to be as low as 10 per cm^3 above the layer and their temperature to be as low as 2000 K, the proton flux into the layer will be a few times 10^6 per cm^2 and sec. The positive-ion fluxes measured below the layer have seldom surpassed 10^6 per cm^2 sec

Block and Fälthammar (1968) have considered in some detail what the effect would be in the upper ionosphere of electric currents along the magnetic field lines.

It is evident that if an electric field associated with the temperature gradient is the cause of the observed field-aligned ion fluxes, a field-aligned anisotropy in the flux of positive ions will be seen only when there is an energetic electron flux into the ionosphere from the magnetospheric plasma. The ESRO 1 particle data which have been investigated hitherto all show this association (Hultqvist *et al.*, 1970).

The field-aligned ions accelerated in the transition layer will be of ionospheric origin. It is known that over the polar cap O$^+$ ions are sometimes the predominating ion species even at 3000 km altitude (Hoffman, 1968). If the layer were located at altitudes below several thousand km, one would expect to see appreciable fluxes of O$^+$ and He$^+$ ions among the ions accelerated by the magnetic field-aligned electric field. No measurements of O$^+$ fluxes seem to have been made. Of course, the ions measured by Hultqvist *et al.* (1970) and others may have been O$^+$ ions, but whether they actually were or not, is unknown.

Concerning energetic helium atoms and ions there are some conflicting results from spectrophotometric measurements on auroral luminosity. While Eather (1968) failed to detect any helium emissions, Stoffregen (1969) measured transient He i emissions during the break-up phase of two strong auroral events. It is, however, likely that these measurements should be made at higher latitudes than has been the case hitherto in order to see these emissions more frequently. The precipitated helium ions which produce these emissions may possibly have their origin in the solar wind, so only if the fraction of the He$^+$ flux were very large, or if the flux were shown to be field aligned, could one be sure that ionospheric helium was measured. The most decisive measurement in favour of the transition layer origin would of course be one of appreciable fluxes of energetic O$^+$ ions.

The potential difference over the transition layer is expected to increase with increasing electron equivalent temperature in the plasma sheet, if diffusion is important. If it is assumed that this equivalent temperature is mostly lower than 10^8 K, as measurements indicate (see e.g. Vasyliunas, 1969), then the anisotropy should not be seen at all or only rarely at say 10 keV. This would explain why field-aligned anisotropy has not been observed in earlier pitchangle measurements above 10 keV.

On the other hand, the real value of the potential drop may depend fairly strongly on the shape of the energy and density distributions and may perhaps surpass the value corresponding to an equivalent high-temperature thermal plasma. As only very few pitchangle measurements have been made in the proton energy range 10–30 keV it is probably not safe to say that field-aligned ions do not occur in this range.

A temperature of 10^8 K may correspond to a potential drop of up to several kilovolts in the transition layer for a quasithermal plasma. This equivalent temperature was needed for the effect to be detectable with the 6 keV positive-ion detectors on the ESRO 1 satellites, and as 10^8 K is above the average equivalent temperature (of the order of 10^7 K; Vasyliunas, 1969) in the plasma sheet, one expects the anisotropy to be detectable only in a minority of the measured cases, as has been seen. On the same basis it is expected that the anisotropy shall occur more frequently in the 1 keV detectors on ESRO 1B. Unfortunately only very little data from this satellite is yet available, so we cannot yet say whether this is true or not.

The energy spectrum of the field-aligned ions is expected to extend from zero up to the energy value corresponding to the total electric potential difference over the layer in the early phase before the transition layer has been emptied of plasma. After this has happened – if it happens – a monoenergetic spectrum is expected, as the cold ions which enter the layer from above have to pass through the entire potential drop on their way down into the ionosphere. Within the transition layer the energy distribution of the field-aligned ions will, of course, be height dependent.

The transition layer will accelerate local cold electrons upwards and will produce an electron flux extremely well collimated along the magnetic field lines far above the layer. If on closed field lines, these electrons will travel to the magnetic conjugate point (unless the beam is broken up by instabilities) and will lose their excess energy again completely in the transition layer on the other hemisphere, if the potential differences are identical on the two hemispheres. If there is some difference between them, field-aligned electron fluxes with the difference energy will penetrate into the ionosphere on that hemisphere where the potential difference over the layer is smallest.

However, instabilities are likely to occur in such very well collimated electron beams and they may, in fact, be an important source of waves and turbulence all along the magnetic field lines.

Field-aligned electron fluxes have been observed, as is well known. According to the ESRO 1 measurements they do not occur as frequently and with so high degrees of anisotropy as in the case of the positive ions.

The effect of the transition layer on the ion density would be to decrease it above the layer and to increase it in that part of the ionosphere where the accelerated ions are stopped. Within the layer the ion density would be lowered, as mentioned before. The light ions will be transported downward faster than the heavier ones, so one would expect that the relative O^+ density would increase in the region where the electric field exists until the transition layer has been emptied of thermal plasma. After that, protons diffusing from above should dominate. The observations of extremely-low-plasma densities at a few thousand km altitude reported by Hagg (1967) and others may possibly be due to persistent electric fields of the kind dealt with in this report.

An alternative cause of the low-plasma density is, of course, the polar wind. A transition layer of several hundred or even thousands of volts potential drop will strongly influence a polar wind. It will, in fact, prohibit the outflow of thermal positive ions completely, so it is obvious that a transition layer with the electric field pointing downward and a polar wind cannot exist at the same time in any one region.

8. Concluding Remarks

The following are some crucial measurements to be made in order to prove or disprove the existence of and to investigate the properties of the kind of transition layer described above:

– Measurements of energetic O^+ ions should be made.

– The electric field of the layer ought to be measured by means of an electric-field experiment on a rocket launched along a magnetic field line if possible. The rocket has to reach altitudes of several thousand km.

– One should also be able to measure in the equatorial plane electrons filling up only a very small part of the loss cone (unless the electron beam becomes unstable and breaks up, creating waves and turbulence).

It seems more evident than ever that the auroral particle acceleration process is not a well localized mechanism, but is distributed over the whole magnetosphere, from the distant tail to the upper ionosphere.

Acknowledgements

The work, on which this paper is based has been sponsored in part by the Swedish Natural Science Research Council and the Swedish Board for Technical Development. The satellites ESRO 1A, B were built and operated by ESRO and launched by NASA.

References

Akasofu, S.-I.: 1964, *Planetary Space Sci.* **12**, 273.
Akasofu, S.-I.: 1965, *Space Sci. Rev.* **4**, 498.
Akasofu, S.-I.: 1966a, *Space Sci. Rev.* **6**, 21.
Akasofu, S.-I.: 1966b, *Planetary Space Sci.* **14**, 587.
Akasofu, S.-I.: 1968, *Polar and Magnetospheric Substorms*, D. Reidel, Dordrecht, Holland.
Alfvén, H.: 1939, *Kgl. Sv. Vetenskapsakad. Handl.* III, 18, No. 3, Stockholm.
Alfvén, H.: 1940, *Kgl. Sv. Vetenskapsakad. Handl.* III, 18, No. 9, Stockholm.
Alfvén, H.: 1958, *Tellus* **10**, 104.
Alfvén, H. and Fälthammar, C.-G.: 1963, *Cosmical Electrodynamics*, Clarendon Press, Oxford.
Angerami, J. J. and Carpenter, D. L.: 1966, *J. Geophys. Res.* **71**, 711.

Axford, W. I.: 1968, *J. Geophys. Res.* **73**, 7855.
Axford, W. I. and Hines, C. O.: 1961, *Can. J. Phys.* **39**, 1433.
Bame, S. J.: 1968, in *Earth's Particles and Fields* (ed. by B. M. McCormac), Reinhold, New York, p. 373.
Bame, S. J., Asbridge, R., Felthauser, H. E., Hones, E. W., and Strong, I. B.: 1967, *J. Geophys. Res.* **72**, 113.
Banks, P. M. and Holzer, T. E.: 1968, *J. Geophys. Res.* **73**, 6846.
Banks, P. M. and Holzer, T. E.: 1969a, *J. Geophys. Res.* **74**, 3734.
Banks, P. M. and Holzer, T. E.: 1969b, *J. Geophys. Res.* **74**, 6304.
Banks, P. M. and Holzer, T. E.: 1969c, *J. Geophys. Res.* **74**, 6317.
Bauer, S. J.: 1969, *Proc. IEEE* **57**, 1114.
Bezrukikh, V. V. and Gringauz, K. I.: 1966, *Space Research*, (ed. by G. A. Skuridin *et al.*), NASA Techn. Translat. TTF-389, May 1966.
Birkeland, Kr.: 1908, *The Norwegian Aurora Polaris Expedition* 1902–1903, Vol. I, Christiania 1908, Section 1.
Birkeland, Kr.: 1913, *The Norwegian Aurora Polaris Expedition* 1902–1903, Vol. 1, Christiania 1913, Section 2.
Block, L. P. and Fälthammar, C.-G.: 1968, *J. Geophys. Res.* **73**, 4807.
Boström, R.: 1968, *Ann. Geophys.* **24**, 681.
Brice, N.: 1967, *J. Geophys. Res.* **72**, 5193.
Buneman, O.: 1959, *Phys. Rev.* **115**, 503.
Burch, J. L.: 1968, *J. Geophys. Res.* **73**, 3585.
Carpenter, D. L.: 1962, *J. Geophys. Res.* **67**, 135.
Carpenter, D. L.: 1966, *J. Geophys. Res.* **71**, 693.
Catchpoole, J. R.: 1969, *Australian J. Phys.* **22**, 733.
Chapman, S. and Ferraro, V. C. A.: 1931, *Terrest. Magnet. Atmosph. Elec.* **36**, 77, 171.
Chapman, S. and Ferraro, V. C. A.: 1932, *Terrest. Magnet. Atmosph. Elec.* **37**, 141, 421.
Chapman, S. and Ferraro, V. C. A.: 1933, *Terrest. Magnet. Atmosph. Elec.* **38**, 79.
Dayharsh, T. I. and Farley, W. W.: 1965, *J. Geophys. Res.* **70**, 5361.
Derblom, H.: 1968, *Ann. Geophys.* **24**, 163.
Eather, R. H.: 1968, *Ann. Geophys.* **24**, 111.
Eviatar, A., Lenchek, A. M., and Singer, S. F.: 1964, *Phys. Fluids* **7**, 1775.
Feldstein, Y. I.: 1960, *Investigations of the Aurora*, Acad. Sci. USSR vol. 4, p. 61.
Feldstein, Y. I.: 1963, *Geomagnetizm i Aeronomiya* **3**, 183.
Feldstein, Y. I.: 1964, *Tellus* **16**, 258.
Feldstein, Y. I.: 1966, *Planetary Space Sci.* **14**, 121.
Feldstein, Y. I. and Solomatina, E. K.: 1961, *Geomagnetizm i Aeronomiya* **1**, 475.
Field, E. C. and Fried, B. D.: 1964, *Phys. Fluids* **7**, 1937.
Finkelnburg, W. and Maecker, H.: 1956, *Encyclopedia of Physics* (ed. S. Flügge), Vol. XXII, p. 254.
Frank, L. A.: 1967a, *J. Geophys. Res.* **72**, 1705.
Frank, L. A.: 1967b, *J. Geophys. Res.* **72**, 3753.
Frank, L. A.: 1968a, in *Physics of the Magnetosphere* (ed. by R. L. Carovillano, J. F. McClay, and H. R. Radoski), D. Reidel, Dordrecht, Holland, p. 271.
Frank, L. A.: 1968b, in *Earth's Particles and Fields* (ed. by B. M. McCormac), Reinhold, New York, p. 67.
Gringauz, K. I.: 1961, *Space Res.* **2**, 539.
Gringauz, K. I., Bezrukikh, V. V., Ozerov, V. D., and Rypchinsky, R. Ye.: 1960a, *Soviet Phys. Doklady* **5**, 361.
Gringauz, K. I., Kurt, B. G., Moroz, V. I., and Shklovsky, I. S.: 1960b, *Soviet Astron.–AJ* **4**, 680.
Hagg, E. L.: 1967, *Can. J. Phys.* **45**, 27.
Hoffman, J. H.: 1968, *Trans. Am. Geophys. Union* **49**, 253.
Hoffman, J. H.: 1969a, *Proc. IEEE* **57**, 1063.
Hoffman, R. A.: 1969b, Paper presented at the NATO Advanced Study Institute, Tretten, Norway, March 1969.
Hoffman, R. A.: 1969c, *J. Geophys. Res.* **74**, 2425.
Hones, E. W., Jr.: 1968, in *Physics of the Magnetosphere* (ed. by R. L. Carovillano, J. F. McClay, and H. R. Radoski), D. Reidel, Dordrecht, Holland, p. 392.

Hones, E. W., Jr.: 1969, in *Aurora and Airglow* (ed. by B. M. McCormac and A. Omholt), Reinhold, New York, p. 351.

Hones, E. W., Jr., Singer, S., and Rao, C. S. R.: 1968, *J. Geophys. Res.* **73**, 7339.

Hultqvist, B., Borg, H., Riedler, W., and Christophersen, P.: 1970, *Planetary Space Sci.*, in the press.

Johansen, O. E. and Omholt, A.: 1963, *Planetary Space Sci.* **11**, 1223.

Karlsson, E.: 1970, personal communication.

Kennel, C. F.: 1969, *Rev. Geophys.* **7**, 379.

Kennel, C. F., and Petschek, H. E.: 1966, *J. Geophys. Res.* **71**, 1.

Khorosheva, O. V.: 1961, *Geomagnetizm i Aeronomiya* **1**, 615.

Khorosheva, O. V.: 1962, *Geomagnetizm i Aeronomiya* **2**, 696.

Liszka, L.: 1967, *J. Atmospheric Terrest. Phys.* **29**, 1243.

Maehlum, B. N.: 1968, *J. Geophys. Res.* **73**, 3459.

Mende, S. B.: 1968, *J. Geophys. Res.* **73**, 991.

Miller, J. R. and Shepherd, G. G.: 1968, *Ann. Geophys.* **24**, 359.

Mozer, F. S. and Bruston, P.: 1966, *J. Geophys. Res.* **71**, 4461.

Muldrew, D. B.: 1965, *J. Geophys. Res.* **70**, 2635.

Nelms, G. L. and Lockwood, G. E. K.: 1967, *Space Res.* **7**, 604.

Nishida, A.: 1966, *J. Geophys. Res.* **71**, 5669.

Ossakov, S. L.: 1968, *J. Geophys. Res.* **73**, 6366.

Persson, H.: 1963, *Phys. Fluids* **6**, 1756.

Petschek, H. E. and Kennel, C. F.: 1966, *Trans. Am. Geophys. Union* **47**, 137.

Reber, G. and Ellis, G. R.: 1956, *J. Geophys. Res.* **61**, 1.

Reme, H.: 1969, Thesis, CESR, Toulouse.

Retzler, J. and Simpson, J. A.: 1969, *J. Geophys. Res.* **74**, 2149.

Riedler, W.: 1969, in *Intercorrelated Satellite Observations Related to Solar Events* (ed. by V. Manno and D. E. Page), D. Reidel, Dordrecht, Holland, p. 557.

Riedler, W., Hultqvist, B., and Borg, H.: 1970, *Proc. COSPAR Meeting 1970.*

Schield, M. A. and Frank, L. A.: 1969, Paper presented at the Conference on Electric Fields in the Magnetosphere, Rice Univ., Houston, Texas, March 1969.

Sharp, R. D., Johnson, R. G., Shea, M. F., and Shook, G. B.: 1967, *J. Geophys. Res.* **72**, 227.

Stenflo, L.: 1968, *Plasma Phys.* **10**, 551.

Stenflo, L.: 1969a, *Plasma Phys.* **11**, 71.

Stenflo, L.: 1969b, *J. Atmospheric Terrest. Phys.* **31**, 197.

Stenflo, L.: 1970, *J. Plasma Phys.* **4**, 145.

Stoffregen, W.: 1969, *Planetary Space Sci.* **17**, 1927.

Stoffregen, W., and Derblom, H.: 1962, *Planetary Space Sci.* **9**, 711.

Swift, D. L.: 1965, *J. Geophys. Res.* **70**, 3061.

Thomas, J. O. and Sader, A. Y.: 1963, Techn. Rept. No. 6, NASA Grant NsG 30-60, Stanford Electronics Lab., Stanford University.

Thomas, J. O. and Sader, A. Y.: 1964, *J. Geophys. Res.* **69**, 4561.

Thomas, J. O. and Andrews, M. K.: 1968, *J. Geophys. Res.* **73**, 7407.

Timleck, P. L. and Nelms, G. L.: 1969, *Proc. IEEE* **57**, 1164.

Vasyliunas, V. M.: 1968, *J. Geophys. Res.* **73**, 2839.

Vasyliunas, V. M.: 1969, in Proc. of the NATO Advanced Study Institute on the Production and Maintenance of the Polar Ionosphere, Tretten, Norway, April 1969.

Vernov, S. N., Melkinov, V. V., Savenko, I. A., and Savin, B. I.: 1966, *Space Res.* **6**, 746.

Warren, E. S.: 1963, *Nature* **197**, 636.

Wax, R. L. and Bernstein, W.: 1970, *J. Geophys. Res.* **75**, 783.

Whalen, B. A., McDiarmid, I. B., and Budzinski, E. E.: 1967, *Can. J. Phys.* **45**, 3247.

Wiens, R.: 1968, Ph.D. Thesis, Inst. Space and Atmosph. Studies, Univ. Saskatchewan, Saskatoon.

THEORETICAL MODEL OF F-REGION STORMS

TATSUZO OBAYASHI

Institute of Space and Aeronautical Science, University of Tokyo

and

NOBUO MATUURA

Radio Research Laboratories, Kokubunji, Tokyo, Japan

Abstract. Large-scale F-region disturbances, which are manifested by changes in the global electron-density profile, are known to be accompanies by magnetic storms. The electron density in the F region is generally enhanced in low latitudes, while it is depressed in middle and high latitudes. Several mechanisms are considered to contribute to the electron-density change, and it is shown that the abundance ratio of atomic oxygen to molecular oxygen or nitrogen is a dominant parameter to control the electron-density change. A theoretical model of F-region storms, which is compatible with the observed features, is proposed.

1. Introduction

Ionospheric F-region disturbances, commonly called ionospheric storms, are characterized by changes in the global electron-density profile over the period of a few days accompanied by geomagnetic storms. Extensive studies on morphology of ionospheric storms were made by using ground-based ionospheric data, and some statistical features of ionospheric storms have been well established for the bottomside ionosphere, especially, for the peak electron density of the F2 layer (Reviews: Obayashi, 1964; Rishbeth, 1968).

Recently, the data on topside ionosphere are obtained by plasma probes and beacons on rockets and satellites, by incoherent backscatter radars and by topside sounders. New information on disturbances in the topside ionosphere are now available, though the data are still in somewhat limited amount as compared with those of the bottomside ionosphere. The morphology of the disturbance effect on the topside ionosphere has been studied by several workers (Summary papers were given by Warren, 1969; Chan and Colin, 1969).

The storm variations of the electron density manifest complicated features depending on stage of storm development, location (altitude, latitude and longitude), season, local time and so on. Polar F-region disturbances are especially complicated and violent, and probably various parameters may take part in agitation of the polar F-region. Interactions between ionosphere and magnetosphere contribute to the polar ionosphere to a great extent and their processes are being brought to light by topside experiments.

In the present report a typical morphology of ionospheric storms will be described with a simplification of the complicated phenomena but with an emphasis on essential points of the features. Several fundamental processes yielding the disturbance effects on the electron density of the F-region ionosphere are discussed, and a theoretical

Dyer (ed.), Solar-Terrestrial Physics/1970: Part IV, 199–211. All Rights Reserved.

model of ionospheric storms, which is compatible with the observed features, will be proposed.

2. General Morphology of F-Region Storms

Ionospheric storms are generally accompanied by geomagnetic storms and appear as electron density changes from normal to disturbed conditions on a world-wide basis.

Onset of an ionospheric storm is generally difficult to identify, unlike the sudden commencement of geomagnetic storm. This may partly be due to the intermittence of data acquisition by ionosonde and partly be due to different storm generation mechanisms. A large electron-density variation was observed during the initial phase of the exceptionally severe magnetic storm on February 11, 1958 (Matuura, 1963), however, general characteristics during the initial phase of storms have not yet been established.

During the main phase of storms there are marked changes in electron-density profile. A typical average feature of the electron-density variation will be described on the basis of statistical studies and it is also shown in Figure 1 including both bottomside and topside data of the ionosphere.

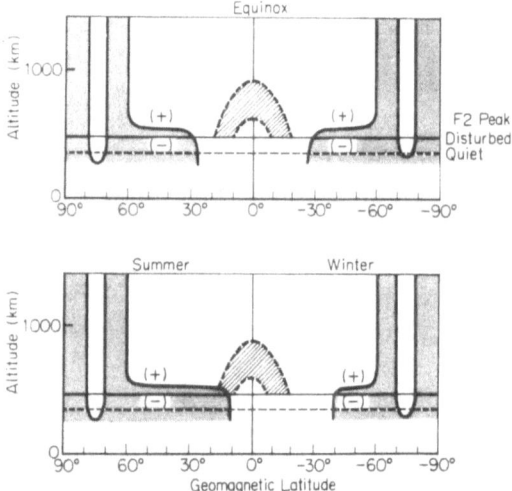

Fig. 1. The distribution of electron density variations of the ionosphere during the main phase of a geomagnetic storm (daytime).

At moderate latitudes, maximum electron density of F2 peak, NmF2, shows a remarkable seasonal dependence in its storm variations, and the region where ΔNmF2$>$0 tends to be accentuated in winter. At low latitudes including the equator, NmF2 is increased at all seasons. Geomagnetic latitudes at the transition between two regions where ΔNmF2$>$0 and where ΔNmF2$<$0 are based on the statistical study by Lange-Hesse (1965). At high latitudes, changes of NmF2 are violent and full of variety in space and time. The overall storm effect can be represented as a decrease of NmF2, but near the auroral belt an increase of NmF2 is also seen.

Storm effect on the topside ionosphere at middle and low latitudes including the equator is a general increase of electron density (Reddy *et al.*, 1967; Ondoh, 1967; Sato, 1968). Electron-density reduction in a dome-shaped region over the equator is recognized during disturbed period in daytime (Sato, 1968). At high latitudes, storm variations of the topside electron-density profile are highly complex and it is difficult to generalize them in a simple picture. However, enhancement of electron density at the belt roughly along the auroral oval and depression of electron density outside the belt may be a possible feature (Nishida, 1967; Sato and Chan, 1969). Interesting observational results are obtained from experiments on Alouette satellites (Norton and Findlay, 1969), and they are reproduced in Figure 2. The results indicate that the electron temperature is enhanced and that the electron density is depressed along the magnetic field lines which intersect the middle latitude red arc. They also show that both the electron temperature and electron density are enhanced over the region of active auroras.

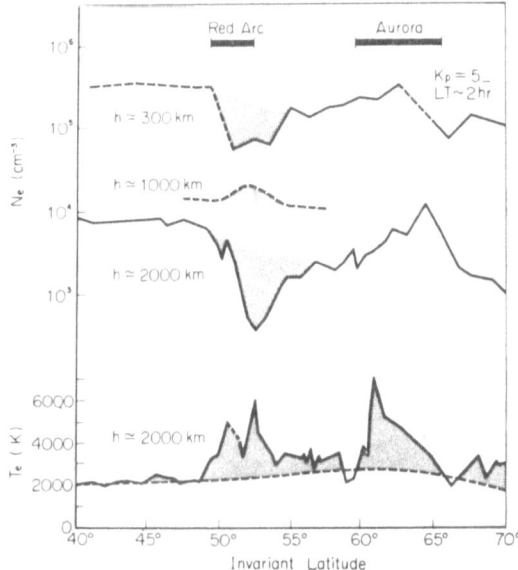

Fig. 2. High latitude ionospheric phenomena associated with a red-arc and auroras (Alouette II, 0900–0915, 29 Sept. 1967; observational data are due to Norton and Findlay, 1969).

Variations of F2 peak height during storms are characterized by a general increase throughout all seasons and all latitudes (Matsushita, 1963), and the amount of the increase is the order of 100 km. However, the height of maximum electron density is remarkably lowered during the *G*-condition ($f_0F2 < f_0F1$) which is sometimes observed at high latitudes in disturbed period (Herzberg and Nelms, 1969). A general increase of slab thickness of the F2 layer during a storm is also recognized both from bottomside N(h) profiles and from Faraday rotation experiments (Taylor and Earnshaw, 1969).

Increases of thermospheric neutral gas temperature (Jacchia *et al.*, 1967) and charged particle temperature, especially electron temperature (Evans, 1966), are important parameters in relation to the storm effect on the ionosphere, and so are the geomagnetic variations. The temperature increment with respect to the degree of geomagnetic agitation K_p at high latitudes is larger by a factor of 2–3 than that at low latitudes, which is demonstrated in Figure 3 (Roemer, 1969). Thus, it is evident that global changes in the geomagnetic field, ionospheric electron density and temperature are three major manifestations of the disturbance taking place in the earth's upper atmosphere.

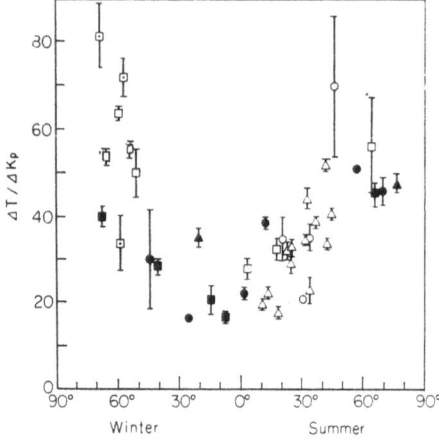

Fig. 3.   Thermospheric temperature increments $\Delta T/\Delta K_p$ during geomagnetic disturbances as a function of latitude for winter and summer conditions (Roemer, 1969).

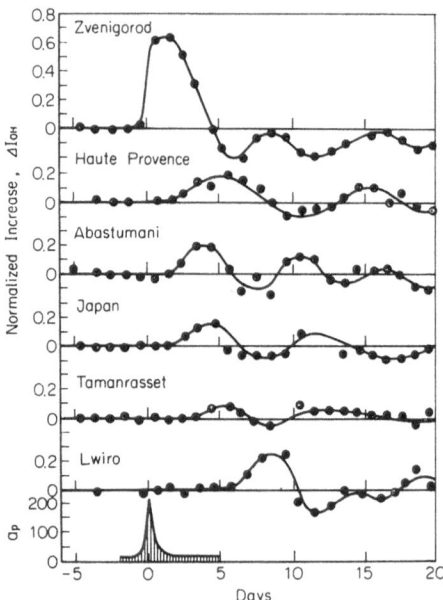

Fig. 4.   Average storm-time changes in the intensity of OH emission during geomagnetic storms at different latitudes (Shefov, 1969).

Other global influences are found in auroral activity, travelling ionospheric disturbances TID, and hydroxyl airglow emission of the upper atmosphere. It has been suggested that TID's are caused by atmospheric gravity waves which are generated in the polar region during disturbed condition and propagate to lower latitudes (Davis and da Rosa, 1969). Fluctuations in intensity and rotational temperature of OH airglow emission have been detected during the period of geomagnetic storms (Shefov, 1969). As is shown in Figure 4, an enhancement of the emission occurs immediately after the commencement of geomagnetic storms at high latitudes, and the effect spreads toward lower latitudes with an appreciable time-delay.

3. Fundamental Processes for Electron-Density Variations

The changes of the electron density in the ionosphere are determined by the continuity equation,

$$\partial N/\partial t = Q - L - \text{div}(NV), \tag{1}$$

where N is the electron density, Q and L the rates of electron production and electron loss per unit volume, respectively, and V the effective velocity of the electron gas which may arise from three components as follows,

$$V = V_D + V_N + V_E, \tag{2}$$

where V_D is due to the effect of ambipolar diffusion, V_N due to the effect of air drag, and V_E due to the effect of electromagnetic drift. Among the three, V_D is important for the F2 layer formation (Yonezawa, 1956), and V_E is known to be dominant at the equatorial region (Bramley and Peart, 1965; Moffett and Hanson, 1965). The effects of V_N upon the electron-density distribution in the F region were suggested (Kohl *et al.*, 1968).

Several fundamental processes suggested to interpret the changes in the electron density during storms will be summarized and discussed below.

3.1. PARTICLE PRECIPITATION

Effects of precipitating particles are important in the polar region through ionization, heating and inflow of hot plasma into the F region. Enhancement of the electron density at high latitudes, as seen in Figures 1 and 2, and also phenomena of G'-condition which arise from electron density enhancements in the F1 and E regions (Herzberg and Nelms, 1969) may be produced by precipitating particles.

3.2. IONOSPHERIC HEATING

It is known that the thermospheric temperature and charged particle temperature increase during geomagnetic disturbances. Direct effects of the temperature increase are taken into consideration in the following processes to realize the electron-density variations.

1. *Thermal Expansion*

The elevated F2-peak height and increased slab thickness of the F2 layer during storms indicate the effect of thermal expansion of the ionosphere. Influences of thermal expansion of the ionosphere including the neutral atmosphere were argued in relation to the ionospheric heating during storms (Matuura, 1963; Garriott and Rishbeth, 1963), but the effect arising only from thermal expansion is too small to explain the observed decrease of NmF2 (Thomas, 1966). General increase of the electron density in the topside ionosphere during storms may be interpreted by the effect of thermal expansion at least for the region where O^+ ions are dominant.

2. *Temperature-Sensitive Aero-Chemical Coefficients*

The loss of electrons in the F region is mainly due to the dissociative recombination with molecular ions. For this reason, chemical reactions $O^+ + N_2 \rightarrow NO^+ + N$..... (reaction rate κ_1) and $O^+ + O_2 \rightarrow O_2^+ + O$..... (reaction rate κ_2), which lead to the formation of NO^+ and O_2^+ from atomic ions, play an important part to diminish the electron density. The electron density depression was interpreted by an enhanced loss rate caused by increases of the chemical coefficients accompanied by the atmospheric heating (Yonezawa, 1963). Some results from laboratory experiments on temperature dependence of those chemical coefficients were reported. Schmeltekopf *et al.* (1967) showed an increase of κ_1 by orders of about 2 when N_2 vibrational temperature T_v increases from 300 K to 6000 K. However, theoretical study by Stubbe (1969) showed that κ_1 sensitivity on T_v i s much reduced when N_2 gas temperature is higher than about 1000 K. The coefficient κ_2 increases almost linearly with O_2 gas temperature above about 650 K, but the estimated increase of κ_2 is less than factor of 2 when O_2 gas temperature increases from 650 K to 2500 K.

3.3. CHANGES IN ATMOSPHERIC COMPOSITION AT THE TURBOPAUSE

The electron density profile in the F region varies with changes of the atmospheric composition which are defined by the changes in $[O]/[O_2]$ and $[O]/[N_2]$ at the turbopause. Relative increase of molecular components will enhance the loss rate of the electron density in the F region, and hence reduce the electron density. Conversely, relative increase of atomic oxygen will enhance the electron density in the F region. They are demonstrated in Figure 5, which shows theoretically computed electron density profiles for different composition ratios: Profile (a) represents the case when $[O]/[O_2, N_2]$ is decreased from the normal condition, and (b) is the reversed case (curves are quoted from Chandra and Herman (1969)).

Changes of the atmospheric composition at the turbopause during disturbed condition have not yet been observed directly. As the possible mechanisms causing changes of the atmospheric composition at the turbopause, an enhancement of atmospheric mixing near turbopause was suggested by Seaton (1956), and Duncan (1969) proposed the effect of global upper atmospheric circulation induced by storm heating.

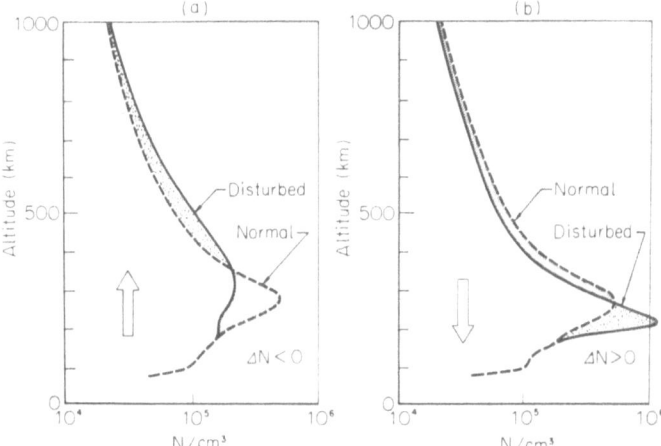

Fig. 5. Electron density profiles for different composition ratios [O]/[O₂, N₂] at the turbopause, (a) relative increase of [O₂, N₂] – upward air motion, (b) relative increase of [O] – downward air motion.

3.4. PLASMA MOTION IN THE IONOSPHERE

1. *Electromagnetic Drift*

Additional electric fields produced during a disturbed period will change the motion of ionospheric plasma by electromagnetic drifts. Interpretations of the disturbance effects on the electron density by the electromagnetic drifts were attempted (Maeda and Sato, 1959). However, it was noticed that the vertical component of electromagnetic drift in the middle latitudes is reduced by the effect of air motion caused by ion drag for the variations of which time scale is the order of a day or longer (Kohl, 1963). Electromagnetic drifts may be effective at the equator where the neutral atmosphere is not moved by ion drag, or otherwise they may be effective for comparatively rapid variations at middle and high latitudes (Rüster, 1969).

2. *Effect of Neutral Winds*

Enhancements of thermospheric temperature accompanied with magnetic disturbances are observed preferentially in high latitudes. Accordingly, an additional atmospheric motion will be induced, being directed towards the equator in both hemispheres. The effect of air drag by such a wind system may cause the changes of the electron density in the F region (Kohl and King, 1967). However, the equatorward wind may raise the F2 layer by air drag in middle latitudes and this will result in an increase of the peak electron density because the loss rate decreases with height (Yonezawa, 1958). Therefore, the effect of air drag alone may not be sufficient to interpret the ionospheric storms.

3. *Plasma Escaping from High Latitude Ionosphere*

An effect of plasma escaping from the ionosphere into the geomagnetic tail has been discussed in order to interpret the formation of the plasmapause and the related trough of the ionosphere (Nishida, 1966; Banks and Holzer, 1968). It is known that the invariant latitudes of the plasmapause and trough shift to the lower latitudes during disturbed condition, resulting in an expansion of the area of depressed electron density (Rycroft and Thomas, 1970).

4. Theoretical Model of Ionospheric Storms

4.1. MIDDLE AND LOW LATITUDES

Precipitating particles may not contribute much to the electron-density variations in middle and low latitudes. Interaction between the changes of the geomagnetic field and the electron density has been studied on the basis of the electromagnetic drift theory. As discussed in the previous section, the effect of electromagnetic drift is not an important factor for the electron-density variations at middle and low latitudes, except for the equatorial region.

Electron-density variations due to the changes in the upper atmospheric temperature have been studied from two different standpoints; one is based on the effect of transport of ionization and the other on the changes of loss rate of the electron density. Transport of ionization arises from the thermal expansion caused by the ionospheric heating and also from air drag by the additional equatorward wind of the neutral atmosphere following the preferential heating at high latitudes. Both of these effects raise the F2 layer upward and cause an increase of the electron density in the topside ionosphere. Thermal expansion results in a depression of NmF2, but the depression is too small to explain the observed amount. Effects of air drag by the equatorward wind may enhance NmF2 at middle latitudes, which is contrary to observation.

Enhanced loss rate of the electron density arises from an increase of the coefficients of the ion-atom interchange/charge-transfer reactions if the coefficients are temperature-sensitive. Temperature dependence of the coefficients has not been applied successfully to the storm-time electron-density variations. Even if the positive temperature dependence is realized, another factor may be necessary in order to interpret the increase of electron density in winter middle latitudes.

Electron loss rate is also affected by the changes in atmospheric composition at the turbopause. Relative increase of molecular components enhances the electron loss rate and relative increase of atomic oxygen reduce the electron loss rate. As possible mechanisms causing the changes in atmospheric composition at the turbopause during storms, the following two may be postulated. One is such that atmospheric mixing near the turbopause is enhanced during disturbed condition by turbulent action, for example by atmospheric gravity waves which are generated at the auroral zone and which propagate to lower latitudes. An enhanced atmospheric mixing increases the relative abundance of molecular components in the F region. However, a difficulty is met by this mechanism because the increase of electron density in winter middle latitudes and in the equator cannot be explained.

The other is such that meridional circulations set up during storms may alter the atmospheric composition at the turbopause. The meridional circulations are generated in both hemispheres by preferential heating of the thermosphere at high latitudes and also by accumulation effect of equatorward Lorenz force ($J \times B$) arising from repeated intensive auroral electrojets. The atmospheric circulation is induced in such a way that the air moves up at higher latitudes followed by equatorward motion above 100 km level and that the air moves down at lower latitudes followed by poleward return motion below 100 km, as schematically shown in Figures 6. By this circulation, atomic oxygen is carried out from higher latitudes and brought into lower latitudes.

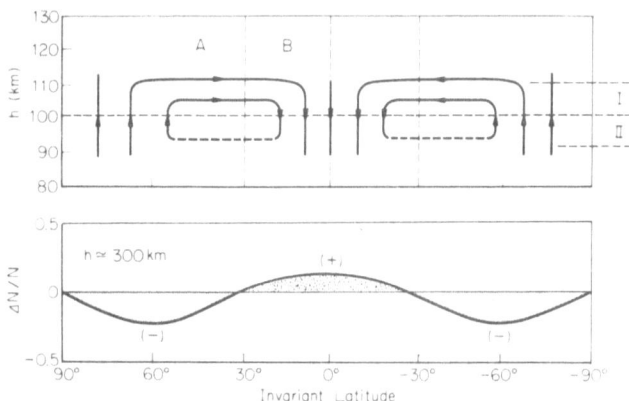

Fig. 6.   Atmospheric circulation caused by heating of polar upper atmosphere, and associated global changes of electron density in the ionosphere.

According to a simple model of the circulation pattern, averaged loss rate of atomic oxygen per unit volume by the divergence effect at the higher latitude region, whose magnitude is equal to that of averaged rate of gain of atomic oxygen per unit volume by the convergence effect at lower latitude region, will be estimated. It is assumed that the circulation pattern is confined to the height range between 90 and 110 km divided into region I and region II by 100 km level and that the circulation pattern is confined to the latitude range between 60° and 0° in each hemisphere which is divided into region A and region B by the latitude of 30°. Continuity of atmospheric particles requires the following equation in a steady state,

$$H_I V_I N_I = H_{II} V_{II} N_{II} \tag{3}$$

where subscripts I and II refer to the values belonging to height region I and II, respectively, H is the height thickness of the regions and is put equal to the scale height of the atmosphere around 100 km, i.e. approximately 10 km, V is the horizontal velocity of the circulation averaged in each region, and N the atmospheric number density, practically the number density of molecular nitrogen averaged in each region. On the other hand, the total net loss of atomic oxygen in the latitude region A, which is balanced to the total net gain of atomic oxygen in the latitude region B, can be represented using Equation (3) as;

$$\text{Net loss in region A} = H_I V_I N_I \left(\frac{[O]_I}{N_I} - \frac{[O]_{II}}{N_{II}} \right) \approx H_I V_I [O]_I, \tag{4}$$

where $[O]$ is the number density of atomic oxygen averaged in each height region and the relation $[O]_I/N_I \gg [O]_{II}/N_{II}$ is used. The averaged loss rate of atomic oxygen per unit volume in region A is estimated as

$$L = \frac{V_I [O]_I}{l} \approx 3 \times 10^6 \text{ cm}^{-3} \text{ sec}^{-1}, \tag{5}$$

where the recombination of atomic oxygen is neglected because the life time of atomic oxygen is longer than about a week for the region above 90 km, and the values of $V_I = 100$ m/s, $[O]_I = 2 \times 10^{11}$ cm^{-3} (Shimazaki, 1967), and the characteristic scale length $l = 6000$ km are used. The values of the loss rate, or the rate of gain, of atomic oxygen given in Equation (5) is a little larger than the production rate of atomic oxygen due to dissociation of molecular oxygen by solar Schumann-Runge continuum at 100 km level (Shimazaki, 1967). Therefore, an appreciable change of atomic oxygen density near 100 km, and consequently in the F region, may be expected.

By those changes of atomic oxygen density, depression of the electron density at high latitudes and enhancement of the electron density at low latitudes will result (Figure 5). Latitudinal asymmetry in the storm changes of the electron density at the solstices as seen in Figure 1 may arise from the effects of the storm generated circulation modulated by background seasonal asymmetry, which is caused by the asymmetry of heat input due to solar EUV radiation (Jacchia, 1965). Johnson (1964) suggested that meridional circulation is generated to compensate the asymmetry in thermospheric heating by solar EUV radiation at the solstices and that the circulation leads to an increase in the relative abundance of atomic oxygen at the F region altitudes in the winter hemisphere, which may probably be the main cause of the winter anomaly in the F region.

The global circulation generated during storms will result in a kind of large scale mixing of the atmosphere and also result in a deep penetration of atomic oxygen into lower atmospheric region where the recombination of atomic oxygen to molecules takes place more quickly. Such an effect will become appreciable about a week later after the commencement of storms. As suggested by Krassovsky (1968), a world-wide enhancement of hydroxyl emission following magnetic disturbances (Shefov, 1969) may arise from the enhanced recombination of atomic oxygen.

4.2. HIGH LATITUDES

The atmospheric circulations which are generated during storms will cause a general depression of the F-region density at high latitudes. Besides, the influence of precipitating particles may be important at high latitudes. Particles, especially electrons, precipitating into the upper atmosphere near the auroral zone ionize the atmospheric particles in D, E, and F regions. By the effect of precipitating particles into the F region, the electron density is enhanced in the F2 layer and also in the topside ionosphere up to high latitudes where O^+ is the dominant ion.

Ionospheric heating at high latitudes may be caused by various factors, such as precipitating particles, Joule heating by auroral electrojets, short period hydromagnetic waves or the conduction of heat from the magnetosphere. The temperature increase following magnetic disturbances appears earlier at high latitudes than at low latitudes by one or two hours, and also the temperature increase at high latitudes is larger than that at low latitudes. These facts suggest that the primary heat input is located in high latitudes. Transport of energy from high to low latitudes may be allotted to the atmospheric circulations or to the atmospheric pressure waves propagating from high to low latitudes.

Ionospheric heating will enhance the escaping of plasma from the ionosphere to the magnetosphere or to the magnetospheric tail, and the electron density in the topside ionosphere will be reduced.

5. Conclusions

The following model is proposed in order to interpret the changes of the electron density during storms (Figure 7).

(i) The polar atmosphere is heated during magnetic storms by precipitating particles, by auroral electrojets, by heat conduction from the magnetosphere, or by hydromagnetic waves.

(ii) As the result of the upper atmospheric heating at high latitudes, atmospheric circulations are generated near the turbopause in both hemispheres such that the air moves up at high latitudes followed by equatorward motion and that the air moves down at low latitudes followed by poleward motion.

(iii) By the influence of the atmospheric circulations, density of atomic oxygen at

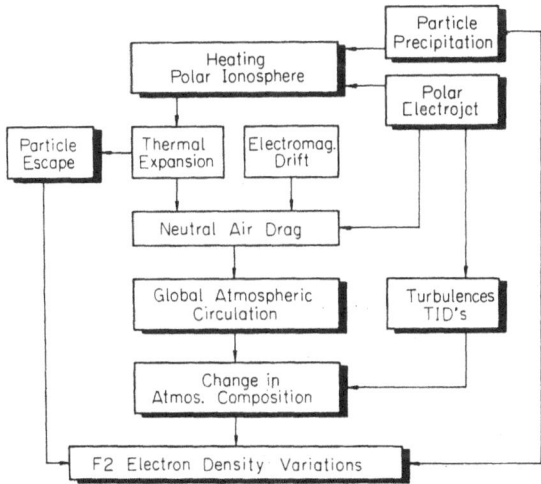

Fig. 7. A schematic model of ionospheric F-region disturbances.

high latitudes is depressed and density of atomic oxygen at low latitudes is enhanced.

(iv) Accordingly, the electron density in the F region decreases at high latitudes and increases at low latitudes. Such a process is shown in Figures 5 and 6.

(v) At solstices, the effects of storm generated circulations are modulated by background asymmetry between both hemispheres, and storm circulation in summer hemisphere is intensified and extended. Asymmetric storm circulation in solstices causes the asymmetric changes in the electron density.

(vi) At high latitudes, general depression of the electron density occurs by the effect of the storm circulations. Besides, precipitating particles enhance the electron density, and short-period electrojets affect the electron density by electromagnetic drifts. Large depression of the electron density is resulted from shifting of the position of trough and plasmapause to lower latitudes which is accompanied by changes in the outer magnetosphere.

(vii) At the equator, there appears a dome-shaped region of electron density depression (daytime) and enhancement (nighttime) caused by electromagnetic drift associated with geomagnetic disturbances in the sense that restrain the equatorial anomaly in normal condition.

In the present review on theoretical model of ionospheric storms, emphasis is placed on the importance of a role played by storm-induced atmospheric circulations, controlling the electron density in the F region. Although the model is consistent in explaining major characteristics of ionospheric storms, further numerical computations of relevant physical processes are required to assess the appropriateness of the present model. It is urged that changes in the atmospheric composition during magnetic disturbances should be observed, experimentally, in order to confirm the theory. For this purpose, new techniques must be developed, capable of measuring directly the atomic oxygen in the E and F regions.

Acknowledgements

We thank Drs A. Nishida and T. Tohmatsu for valuable comments and discussions during the preparation of this review.

References

Banks, P. M. and Holzer, T. E.: 1968, *J. Geophys. Res.* **73**, 6846.
Bramley, E. N. and Peart, M.: 1965, *J. Atmospheric Terrest. Phys.* **27**, 1201.
Chan, K. L. and Colin, L.: 1969, *Proc. IEEE* **57**, 990.
Chandra, S. and Herman, J. R.: 1969, *Planetary Space Sci.* **17**, 841.
Davis, M. J. and da Rosa, A. V.: 1969, *J. Geophys. Res.* **74**, 5721.
Duncan, R. A.: 1969, *J. Atmospheric Terrest. Phys.* **31**, 59.
Evans, J. V.: 1966, *Electro Density Profiles in Ionosphere and Exosphere* (ed. by J. Frihagen), North-Holland Publishing Co., Amsterdam, p. 399.
Garriott, O. K. and Rishbeth, H.: 1963, *Planetary Space Sci.* **11**, 587.
Herzberg, L. and Nelms, G. L.: 1969, *Ann. IQSY* **3**, 426.
Jacchia, L. G.: 1965, *Space Res.* **5**, 1152.
Jacchia, L. G., Slowey, J., and Verniani, F.: 1967, *J. Geophys. Res.* **72**, 1432.

Johnson, F. S.: 1964, Southwest Center for Advanced Studies, Report on Contract Cwb 10531.
Kohl, H.: 1963, *Proc. Intern. Conf. Ionosphere, London*, The Institute of Physics and the Physical Society, p. 198.
Kohl, H. and King, J. W.: 1967, *J. Atmospheric Terrest. Phys.* **29**, 1045.
Kohl, H., King, J. W., and Eccles, D.: 1968, *J. Atmospheric Terrest. Phys.* **30**, 1733.
Krassovsky, V. I.: 1968, *Ann. Geophys.* **24**, 1053.
Lange-Hesse, G.: 1965, *Arch. Elektr. Übertrag.* **19**, 326.
Maeda, K. I. and Sato, T.: 1959, *Proc. IRE* **47**, 232.
Matsushita, S.: 1963, *Proc. Intern. Conf. Ionosphere, London*, The Institute of Physics and the Physical Society, p. 120.
Matuura, N.: 1963, *J. Radio Res. Lab. Japan* **10**, 1.
Moffett, R. J. and Hanson, W. B.: 1965, *Nature* **206**, 705.
Nishida, A.: 1966, *J. Geophys. Res.* **71**, 5669.
Nishida, A.: 1967, *J. Geophys. Res.* **72**, 6051.
Norton, R. B. and Findlay, J. A.: 1969, *Planetary Space Sci.* **17**, 1867.
Obayashi, T.: 1964, *Research in Geophysics*, MIT Press, **1**, 335.
Ondoh, T.: 1967, *J. Radio Res. Lab. Japan* **14**, 267.
Reddy, B. M., Brace, L. H., and Findlay, J. A.: 1967, *J. Geophys. Res.* **72**, 2709.
Rishbeth, H.: 1968, *Rev. Geophys.* **6**, 33.
Roemer, M.: 1969, *Ann. Geophys.* **25**, 419.
Rüster, R.: 1969, *J. Atmospheric Terrest. Phys.* **31**, 765.
Rycroft, M. J. and Thomas, J. O.: 1970, *Planetary Space Sci.* **18**, 65.
Sato, T.: 1968, *J. Geophys. Res.* **73**, 6225.
Sato, T. and Chan, K. L.: 1969, *J. Geophys. Res.* **74**, 2208.
Schmeltekopf, A. L., Fehsenfeld, F. C., Gilman, G. I., and Ferguson, E. E.: 1967, *Planetary Space Sci* **15**, 401.
Seaton, M. J.: 1956, *J. Atmospheric Terrest. Phys.* **8**, 122.
Shefov, N. N.: 1969, *Planetary Space Sci.* **17**, 797.
Shimazaki, T.: 1967, *J. Atmospheric Terrest. Phys.* **29**, 723.
Stubbe, P.: 1969, *Planetary Space Sci.* **17**, 1221.
Taylor, G. N. and Earnshaw, R. D. S.: 1969, *J. Atmospheric Terrest. Phys.* **31**, 211.
Thomas, L.: 1966, *J. Geophys. Res.* **71**, 1357.
Warren, E. S.: 1969, *Proc. IEEE* **57**, 1029.
Yonezawa, T.: 1956, *J. Radio Res. Lab. Japan* **3**, 1.
Yonezawa, T.: 1958, *J. Radio Res. Lab. Japan* **5**, 165.
Yonezawa, T.: 1963, *Proc. Intern. Conf. Ionosphere, London*, The Institute of Physics and the Physical Society, p. 128.